ROUNDABOUT:
THE PHYSICS OF ROTATION
IN THE EVERYDAY WORLD

Readings from
"The Amateur Scientist" *in*
**SCIENTIFIC
AMERICAN**

ROUNDABOUT:
THE PHYSICS OF ROTATION IN THE EVERYDAY WORLD

Jearl Walker

Cleveland State University

W. H. Freeman and Company

New York

Library of Congress Cataloging in Publication Data

Walker, Jearl, 1945–
  Roundabout : readings from the Amateur scientist in
Scientific American.

  Bibliography: p.
  Includes index.
  1. Motion. I. Title
QC125.2.W27   1985   531'.11   85–4358
ISBN 0–7167–1724–7
ISBN 0–7167–1725–5 (pbk.)

Printed in the United States of America

2 3 4 5 6 7 8 KP 4 3 2 1 0 8 9 8 7 6

CONTENTS

PREFACE

This book, a collection of articles from "The Amateur Scientist" department of *Scientific American,* deals with examples of rotation in the everyday world. Rotation is fascinating because in spite of its common occurrence it is difficult to understand. We are not accustomed to it as we are to motion along a straight line, such as the motion of cars along a freeway. Indeed, we are so skilled at straight motion that we survive at freeway speeds and some of us even participate in high-speed races.

Rotating objects seem strange, because their movement is not easy to anticipate. For example, when I spin a top I am always amazed that it stands upright in spite of the gravity tugging at it. Even when it leans over, it does not fall. Instead it precesses, that is, the central axis about which it spins circles around the vertical. Such motion is magic.

Rotation is even more mystifying when I participate in it. In fact, the local amusement parks make their money from those of us willing to pay for the thrill of rotation. The same physics shows up in the surprising twists and turns of ballet and judo. Much of the grace of ballet and the skill of judo depends on training one's body to accept the rules of physics in spite of their abstraction and strange terms. All of these mysteries are discussed in the articles here.

I have investigated a variety of toys, too. When I throw a boomerang so that it returns, I am perplexed that it does not continue to travel away from me as a rock would. When I hit a racquetball with spin, I must train myself to expect a surprising bounce from the wall or ceiling. The same application of spin separates the novice from the professional in a game of pool.

Also included here is an analysis of a strange top that turns itself over and a strange stone that insists on spinning in one direction instead of the other. The rotational dynamics behind these two toys seems to give them a mind of their own. Similar dynamics is responsible for the clatter a spinning coin makes as it falls late in its wobbling.

My collection of articles is only the beginning of studies you might undertake. For that purpose I have included extra references at the end of the bibliography. You might follow the excellent work of Wolfgang Bürger, who has studied superballs, diabolos, gyroscopes, and yo-yos (including some modern yo-yos). The books by David Griffing and Peter Brancazio are treasure chests of ideas about rotation in sports. For example, why does a quarterback put spin on a football when he passes it? How does a baseball pitcher manage to throw a curve ball?

There are many more questions that could be asked. How does a cowboy spin a lasso? How does a hula-hoop stay up? Why is the procedure of taking a bicycle around a turn different from taking a motorcycle around one? How does a cat right itself when it is dropped inverted? How does a person perform complicated gymnastics or springboard diving? How does a child begin and then increase the swinging of a swing on the playground?

The list of unexplored topics is long. My fascination with rotation is so strong that I shall continue articles on it in *Scientific American*. If you study rotation, I would be pleased to hear what you find.

Amusement Park Physics

Thinking about physics while scared to death (on a falling roller coaster)

The rides in an amusement park not only are fun but also demonstrate principles of physics. Among them are rotational dynamics and energy conversion. I have been exploring the rides at Geauga Lake Amusement Park near Cleveland and have found that nearly every ride offers a memorable lesson.

To me the scariest rides at the park are the roller coasters. The Big Dipper is similar to many of the roller coasters that have thrilled passengers for most of this century. The cars are pulled by chain to the top of the highest hill along the track. Released from the chain as the front car begins its descent, the unpowered cars have almost no speed and only a small acceleration. As more cars get onto the downward slope the acceleration increases. It peaks when all the cars are headed downward. The peak value is the product of the acceleration generated by gravity and the sine of the slope of the track. A steeper descent generates a greater acceleration, but packing the coaster with heavier passengers does not.

When the coaster reaches the bottom of the valley and starts up the next hill, there is an instant when the cars are symmetrically distributed in the valley. The acceleration is zero. As more cars ascend, the coaster begins to slow, reaching its lowest speed just as it is symmetrically positioned at the top of the hill.

A roller coaster functions by means of transfers of energy. When the chain hauls the cars to the top of the first hill, it does work on the cars, endowing them with gravitational potential energy, the energy of a body in a gravitational field with respect to the distance of the body from some reference level such as the ground. As the cars descend into the first valley much of the stored energy is transferred into kinetic energy, the energy of motion.

If the loss of energy to friction and air drag is small, the total of the potential and kinetic energies must remain constant throughout the descent and even throughout the rest of the ride. The coaster gains kinetic energy and speed at the expense of potential energy. If the first valley is at ground level, the transfer is complete, and for a moment the coaster has all its energy in the form of kinetic energy.

Without energy losses the coaster could climb any number of hills as high as the one from which it is released (but no higher). To be sure, friction and air drag do remove energy from the coaster, and its total energy content dwindles. It can no longer climb high hills, which is why the last stages of the track consist only of low hills.

The length of a ride on a roller coaster depends on the speed. If the ride is to be fast, the launching hill should be high so that the total energy is large. The rest of the track should be low so that most of the energy remains kinetic.

The choice of a seat on a roller coaster makes a difference in the ride. Some people prefer the front seat because the descent from the launching site presents the pleasingly frightening illusion of falling over the edge of a cliff. Other people prefer the psychological security of the rear seat.

The choice of a seat also determines the forces felt by the passenger. Consider the first descent. The front car starts down slowly because little of the coaster's energy is then kinetic. The speed of the cars increases as an exponential function of time, so that the rear car starts down at a much higher speed than the front car did. Although the passengers in the front car get an unobstructed view of the descent, the passengers in the rear car have a stronger sense of being hurled over the edge.

At the edge one force on the passenger is from the change in the direction of his momentum vector. Initially the vector is horizontal, but soon it points toward the

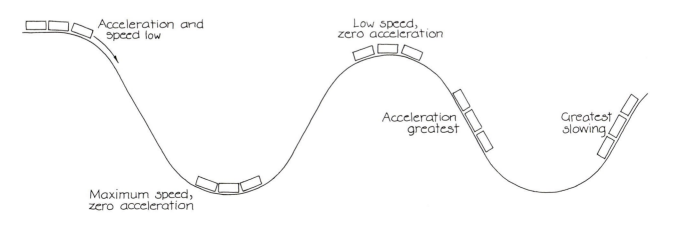

valley. The force necessary to effect this change in direction is delivered by the safety bar or seat belt that keeps the passenger in the car. That force, which points downward and back toward the hill, is part of the thrill of the ride. A passenger in the rear feels the force more than a passenger in the front because the size of the force is proportional to the momentum, which is greater for the passenger in the rear.

The story is different in the valley. Again a force from the coaster is necessary to redirect the passenger's momentum. This time the momentum is initially downward toward the bottom of the valley and then is redirected toward the top of the next hill. The front passenger has a large momentum and is subjected to a large force. By the time the rear car reaches the bottom of the valley the movement of the front cars up the next hill has slowed the coaster. A passenger at the rear has less momentum and is subjected to a smaller force.

At the crest of the hill the passenger gets a force leveling his momentum vector. At the rear of the coaster the force can be considerable if the front is already well down the next slope. To a passenger at the rear who is loosely held in place by a safety bar a fast trip over a hill provides a brief sensation of being lifted from the seat. He arrives at the crest with a large momentum. Until he encounters the safety bar and is redirected he continues to travel upward even though the coaster has leveled out below him. The faster the coaster goes over a hill, the greater the sensation of being thrown free.

The brave passenger is one who rides the roller coaster without holding on. I tried this once while arriving at the crest of a hill at high speed. I avoided being thrown free of the coaster by catching my thighs on the safety bar at the last instant. Thereafter I kept a tight grip on the safety bar.

Roller coasters such as the Big Dipper have been around for more than 50 years. Recently a new type of coaster has appeared. The principles are seen

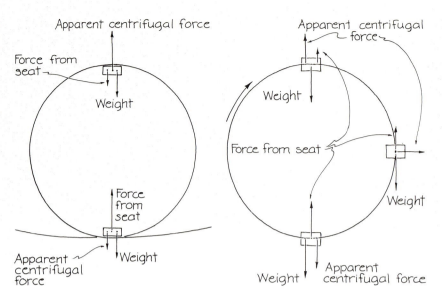

Forces in a coaster that loops   *What you feel in a Ferris wheel*

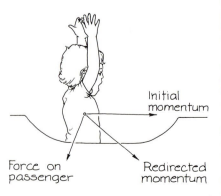

Forces that redirect momentum

in the Double Loop and the Corkscrew. The Double Loop at Geauga Lake begins like the Big Dipper in that a chain pulls the cars to the top of the first and highest hill. After the coaster travels over a few smaller hills (and before it loses too much of its energy to friction and air drag) it runs through two vertical loops. The ride is splendidly unnerving. In the times I managed to open my eyes while traveling through the loops I saw the world turn upside down, the ground race up toward me and the world turn upside down again.

The coaster on the Double Loop is held onto the track by a double set of wheels, one set on the top of the rails and the other set on the bottom. When the coaster is on the normal section of the track, its weight rests on the top set of wheels. When it is in the loop, the other set of wheels can come into play. They keep the cars from flying off the track.

When the coaster enters a loop, I sense three forces. One is my weight, which of course is directed downward. Another is the force from the seat. The third is the apparent centrifugal force downward, which seems to add to my weight; it makes me feel as though I am being pushed into the seat. At the top of a loop the apparent centrifugal force is upward, and I feel light.

The centrifugal force is a fiction. No outwardly directed force is at work. The notion of a centrifugal force is useful, however, since it easily explains what a passenger feels. The perspective of someone on the ground is more to the point: a combination of real forces causes the rider to move in a circle instead of a straight line.

If the circular motion is to be maintained, the net force must be toward the center of the circle. At the bottom of a loop the passenger's weight vector is downward (and therefore away from

the center of the loop). An upward force acts on the passenger from the seat. Since the push from the seat is greater than his weight, the net force points toward the center of the loop and he begins the circular motion. From the passenger's perspective the large push from the seat is sensed as a centrifugal force pressing him into the seat.

At the top of the loop the forces have changed. The passenger's weight is still the same and is still pointed downward, that is, toward the center of the loop. The push from the seat is also downward. The two forces combine in the net force that makes the rider continue in the circle.

This time the force from the seat is smaller. One reason is that at the top of the loop the coaster has less kinetic energy and so is traveling slower. Moreover, the force from the seat is now augmented by the rider's weight vector instead of having to oppose it. The rider senses the force from the seat as a small centrifugal force.

How high must the coaster be at the start of its journey (with essentially no initial speed) if it is to have at the top of the loop the speed that will hold it firmly on the track? To answer the question I made two assumptions. The first was that the coaster has only one car. The second was that the energy losses from friction and air drag are negligible. With these assumptions I found that the first hill must be higher than the top of the loop by at least half the radius of the loop.

The first assumption is a convenient simplification. If the coaster is long, one must consider the rise and fall of its center of mass rather than considering only one car. Since only part of the coaster is at the top of the loop at any given instant, the center of mass never reaches that height, and so less energy is actually required to keep the coaster on the track than would be needed if there were only one car.

As for the second assumption, if the losses of energy were entirely negligible, the unpowered coaster would arrive at the loop with all the energy it got at the launch. The intervening hills and valleys would not matter. They do matter, of course, because they provide more opportunity for energy losses. Therefore the initial hill must be higher than the theory would indicate. On the Double Loop at Geauga Lake the initial hill is considerably higher than the theoretical value, so that the coaster is still traveling at a good clip when it reaches the top of the loop.

The Corkscrew is a similar roller coaster except that the loops are helical. Once the coaster enters the loops it moves in a corkscrew fashion until it emerges again. At two points the passengers are fully upside down.

The physics of this ride is similar to that for the Double Loop. The major difference lies in the direction of the apparent centrifugal force. With the Double Loop the center of the motion in a loop is at a single point. The centrifugal force appears to be directed radially outward from that point. As the coaster travels around the loop, this force rotates in a vertical plane. With the Corkscrew the center of motion continuously

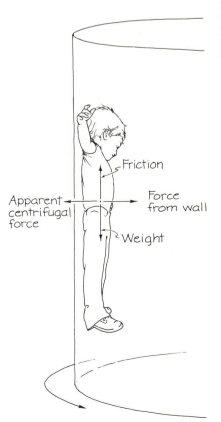

The role of friction in the Rotor

moves vertically and horizontally as the coaster travels through the loops. Hence the direction of the apparent centrifugal force is not confined to a vertical plane. This added feature is one reason the Corkscrew has become so popular with coaster addicts.

Geauga Lake has two other rides that are similar to the standard roller coaster. The water slide starts high above the ground. Water pours down the interior of the slide to provide lubrication and even a small amount of propulsion. The principle is simple: the initial gravitational potential energy is steadily converted into kinetic energy, so that the slider's speed increases during the descent. The lubrication provided by the water diminishes the loss of energy to friction.

The other ride is the Gold Rush Log Flumes. Passengers board a small boat shaped like a hollow log. It is really a car like the ones on the Big Dipper. Water flowing through the flume pushes the boat along until it is engaged by a chain system that drags it up a tall hill. From the crest the boat descends rapidly down a slope of about 45 degrees. At the foot of the hill it speeds into a trough of water, which quickly slows the motion and satisfyingly drenches the passengers. They also seem to be thrown forward, but the experience is illusory; what happens is that they continue to move forward briefly at the former speed.

Most of the other rides at an amusement park are based on rotational mechanics. The mildest of them is the merry-go-round. Here the rate of rotation is just enough to give the passenger a moderate sensation of centrifugal force. He seems to be thrust outward. Actually his body leans outward because the horse moves away from him as it travels in a circle and ends up pulling him along.

The Ferris wheel is similar except that its plane of rotation is vertical. The apparent centrifugal force seems periodically to increase and decrease the passenger's weight. When he passes through the bottom of the circle of travel, the centrifugal force appears to push him downward into the seat as if he then weighed more. In reality the seat pushes strongly against him as it keeps him moving in a circle. This force must be strong because it opposes the passenger's weight. At the top of the circle the passenger has the sensation of being somewhat lighter because the apparent centrifugal force is then upward, seemingly pulling him out of his seat. Actually the sensation comes from the fact that the force from the seat is then smaller.

At midpoint of the descent an even stranger sensation is felt. The force from the seat matches the passenger's weight, and the centrifugal force is outward. Hence the passenger feels as though he is about to be thrown forward out of the compartment.

My favorite among the rotating rides is the Rotor, which is a vertical cylinder with a diameter of about 12 feet. The rider stands with his back against the wall as the cylinder begins to spin. When the maximum spin rate is reached, the floor drops away, but the rider remains stuck to the wall. A particularly agile person might be able to squirm enough to get himself into an angled position or even upside down.

Why does the rider stick to the wall? From his perspective a centrifugal force pins him there. The resulting friction between him and the wall prevents him from falling when the floor is removed. A high rate of spin is called for, so that the apparent centrifugal force generates enough friction.

From the perspective of an outside observer the story is different. The rider is constrained to move in a circle because of a force from the wall. This centripetal force is responsible for the friction. Still, the spin rate has to be high if the force from the wall is to generate enough friction to hold the rider in place.

The Rotor at Geauga Lake has a roughly textured wall to increase the friction. With a smoother wall the centripetal force would have to be stronger to keep the rider from slipping. (One would either have to increase the spin rate or build a cylinder with a larger diameter.) Each time I rode the Rotor I was impressed by the overwhelming sensation that a centrifugal force was pushing me against the wall. In reality the wall was pushing on my back.

In order for the rider to stay in place his weight (a force vector downward) must not exceed the friction (a force vector upward). The amount of friction can at most be equal to the product of the friction coefficient (which depends on the roughness of the surfaces in contact) and the centripetal force from the wall. I estimated that the spin rate had to be about 30 revolutions per minute to hold me against the wall. Indeed, the apparatus did turn at about that rate.

Several other rides at Geauga Lake involve an apparent centrifugal force. The Muzek Express consists of a series of cars moving on a circular track that traverses several small hills. The diameter of the track is roughly 30 feet. The ride is fast, and so the centrifugal force on a passenger is quite strong. The hills provide extra thrills. Usually two people ride side by side in a car. Since they both feel an outward force, the passenger on the outside is squeezed against the wall of the car by the passenger on the inside. The forces are surprisingly large even if a passenger is small. I cannot avoid being pushed into the wall even when the inside passenger is my young daughter, who weighs less than half what I do.

The Enterprise is a rotating ride with cars individually suspended on radial arms extending from a central hub. As the cars begin to move in a horizontal circle the apparent centrifugal force makes the car rotate outward on the radial arm. Soon the car has rotated almost 90 degrees, and the passenger can see the ground directly below the window that originally was on the inside.

This rotation results from the way the mass of the car and the mass of the passenger are distributed with respect to the suspension axis of the car. A combined centrifugal force operates on the common center of mass of the passenger and the car. Initially this point lies below the suspension axis. Also acting through the point is the combined weight of the passenger and the car. These two forces compete in orienting the car. Initially gravity pulls the car into the normal orientation, but as the ride moves faster and the centrifugal force gets stronger the car is rotated increasingly out of the vertical.

This much of the ride was disturbing, but the next part almost did me in. Once the ride had reached its highest speed the large arm that held the central hub was turned to make the plane of the moving cars vertical. I was then moving in a vertical circle, being completely upside down at the top and greatly compressed by the forces acting on me at the bottom. I closed my eyes and began to count the prime numbers.

My next ride also held a surprise. It was a set of swings about 20 feet in diameter suspended from a central hub. When the hub began to turn, I moved in a circle below the rim of the hub. As the speed increased, the apparent centrifugal force moved me outward so that I traveled in a larger circle than before. The faster the hub turned, the larger the circle was.

From my perspective three forces affected me. I still had weight, which was directed downward. The chair and its suspension chains provided a second force directed toward the attachment of the chains to the overhead hub. The third force was the fictitious centrifugal force I felt throwing me outward. The angle between the chains and the vertical was set by the balancing of the three forces. When the speed of the ride increased, the angle also increased, so that the forces again balanced.

The surprise of the ride was that the hub soon tilted about 10 degrees or so out of the horizontal. Part of my travel around the apparatus was then downhill. My speed increased during the descent as potential energy was converted into kinetic energy. As a result I circled the apparatus in a large radius. In the uphill part of the circle I slowed as the hub was forced to hoist me, and so here I circled with a smaller radius.

I ended my busy day at Geauga Lake with three rides that delivered similar types of motion. The first was the

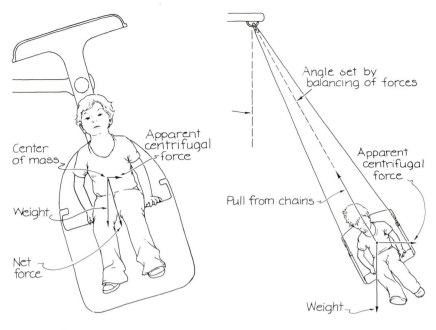

Forces in the Enterprise   *What happens with a rotating swing*

Scrambler, which has long been popular at amusement parks. It consists of a central hub from which several radial arms extend. I call them the primary arms. At the end of each primary arm four secondary arms extend outward. Each one carries at its outer end a car for two or three passengers.

The ride consists of two circular motions. The primary arms rotate steadily about the center of the ride while each set of four secondary arms twirls below the pivot at the end of the primary arm. From an overhead perspective the primary arms move clockwise, the secondary arms counterclockwise. (In a related

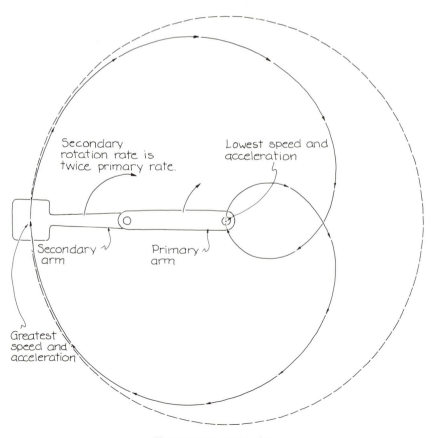

The motions in a Calypso ride

ride called Calypso both motions are clockwise.)

I set about studying the types of motion in rides such as the Scrambler and the Calypso. To model what happens to a rider I focused on a single primary arm (turning clockwise) and a single secondary arm (turning in either direction). As the primary arm makes a full revolution does the passenger loop or spiral? Where are the speed and the acceleration greatest? How should the arms rotate to give an unforgettable ride? Should the arms be approximately the same length (as they are in the Scrambler and the Calypso)?

I found that if the arms are the same size and rotate at the same rate and in the same direction, the ride is bound to be rather boring, because the passenger merely goes in a large circle. The ride is not much better if the arms turn in opposite directions. In this arrangement the passenger would travel on a straight line over the center to the opposite side of the ride and then would return on the same line.

A better ride results when the primary and secondary arms rotate at different rates. Suppose the secondary arm turns twice as fast as the primary one. When the primary and secondary arms move in a clockwise direction, as they do in the Calypso, the passenger first spirals in toward the center of the ride and then out again, so that he travels through a loop on the side opposite to the starting point. After spiraling outward he passes through his initial location and begins the trip again.

His speed and acceleration are highest when he is farthest from the center, that is, when he passes through the initial point. They are lowest when he passes over the center of the ride. My calculations approximate the conditions of the Calypso but are off somewhat because to simplify matters I visualized arms of equal length. In order to accommodate all the primary and secondary arms on the Calypso the secondary arms are shorter than the ones in my calculations, so that the cars do not crash near the center of the ride.

If the primary and secondary arms turn in opposite directions, as they do in the Scrambler, a more interesting motion results. At first the passenger moves counterclockwise, but he quickly heads for the center of the ride and then outward again. When the arms are fully extended, he is directed back toward the center. When the primary arm has completed one revolution, the passenger has traveled through a pattern resembling three narrow petals. He is moving at the highest speed when he passes over the center of the ride. Surprisingly, the acceleration there is the lowest. The speed is the lowest and the acceleration the highest when the passenger is farthest from the center.

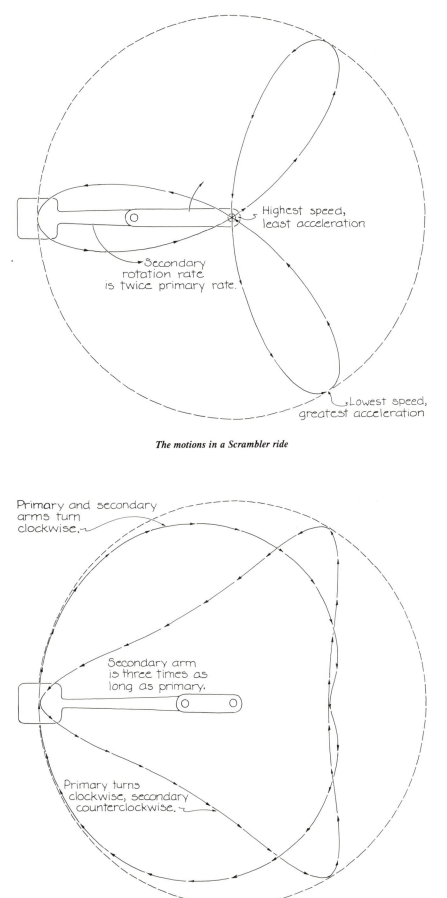

The motions in a Scrambler ride

The motions in a Calypso or a Scrambler with arms of unequal length

The low speed at that point results because the circular motion of the secondary arm is carrying the passenger counterclockwise while the motion of the primary arm is clockwise. The two motions oppose each other when the passenger is farthest from the center and augment each other as he is carried close to the center. The high acceleration at the far point develops because the direction of the velocity out there is changing rapidly.

On my home computer I worked out the paths for other conditions. If the primary arm is much longer than the secondary arm, the passenger might spiral toward and then away from the center. In other cases he would move through a path consisting of a series of cusps or loops superposed on a large circle. Another interesting situation arises when the secondary arm is longer than the primary one. If the secondary arm turns slower than the primary one and in the same direction, the passenger will spiral gradually toward the center and then away from it. If the motions are in opposite directions, the ride will include an abrupt change in direction.

My last ride of the day was on the Tilt-A-Whirl. I sat in a compartment that was free to move around a small circular track, pivoting on a center point at my feet. Six of these compartments, running on a hilly track, moved around the center of the apparatus while they were also pivoting on their individual center points. The result was three types of motion. One was a basic counterclockwise circling about the center of the entire ride. The second was a smaller circling of the compartment in either direction. The third was the vertical motion over the hills.

The interesting part was that I could often control the small circling of my compartment by anticipating the hills and shifting my weight. When the compartment was turning in its small circle and beginning to descend from a hill in the larger motion, I threw my weight in the direction of the compartment's turn. I was transforming some of the potential energy of my body (from being up on the hill) into kinetic energy applied to the rotation of the compartment.

When I timed this exercise correctly, I set the compartment spinning rapidly. My experience was similar to that in the preceding two rides. If the spin direction was in the same direction as the large-scale circling of the ride, the speed and centrifugal force were quite large when I was farthest from the center of the ride. If the spin was in the opposite direction, the acceleration was high when I was far from the center but the speed was high closer to the center.

You might explore the amusement parks near you for other rides. One that I have heard described, but lack the courage to even look at, is Demon Drop. The victim—sorry, the passenger—is secured in a chair that is lifted 131 feet and then dropped in a virtually free fall to the ground. The huge kinetic energy of the ride is apparently dissipated when the apparatus curves into a horizontal section of track at the bottom of the fall. I have no intention whatever of verifying this assumption.

NOTES

Chris Chiaverina, a teacher at Barrington High School in Barrington, Illinois, sent me a manuscript prepared by his class on experiments conducted at Old Chicago Amusement Park. Their work was the stimulus for my article. In addition to investigating some of the rides I have described, they also studied the bumper cars and measured the accelerations of rides with a homemade accelerometer.

To construct such a device, attach a string to the center point of a protractor. On the other end of the string attach a washer to supply weight. Invert the protractor so that its straight edge is horizontal and the taut string passes by the mark for 90 degrees. When the device undergoes acceleration, the string is deflected to one side. The angle of the deflection is a measure of the acceleration: The size of the acceleration is equal to the acceleration of gravity multiplied by the tangent of the deflection. For example, if the string is deflected by 10 degrees, the acceleration is about 0.17 that of gravity.

C. K. Stedman of New Denver, British Columbia, pointed out that the acceleration of a roller coaster is not zero if the valley into which it travels is curved. (I had assumed that the valley is relatively flat until the next hill is reached.) With a curved valley, a rider experiences an apparent centrifugal acceleration directed downward. The size of the acceleration is the square of the speed divided by the radius of curvature of the valley.

Arthur Eisenkraft, who teaches physics at Briarcliff High School in Briarcliff Manor, New York, and Donald Baker Moore of Explosive Technology in Fairfield, California, described a strange perception you might have while riding the Rotor. As the Rotor spins you will feel as though you were leaning backward. The illusion comes from the experience that "down" is the direction in which gravity pulls you. When in the Rotor you have an overwhelming illusion that the centrifugal force is pushing you outward from the center of the Rotor. You will then believe that a single force (the combination of gravity and the centrifugal force) is acting on you. Since this combined force is directed downward and backward, you interpret it as being gravity and then define "down" as being in that direction. The result is a perception that you are leaning backward.

Dag O. Ellingsen of the Ship Research Institute of Norway sent me a description of a lifeboat designed to fall like the Demon Drop. It is intended to be a means to evacuate people from an ocean oil rig during an emergency. The boat has a steel hull and polyurethane foam seats. The boat and its seats are built to protect the occupants during the collision with the water. "So far, the boat has been dropped from 40 meters (131 feet) without persons inside, once from 30 meters (99 feet) with persons inside and several times from 20 to 25 meters (66 to 82 feet) also with persons inside."

J. G. Krol of Anaheim, California, wrote me about a ride that once was at Geauga Lake Amusement Park. "It consisted of two small, roughly cylindrical cars, each dangling from a long vertical arm, rather like a pair of ball-peen hammers hanging heads down side by side. A central pillar supported the two arms at their tops.

"The ride began with the two arms swinging a few degrees in opposite direction, opening up like scissor blades. The pendulum motion then reversed, with greater amplitude. As the pendulum motion gradually built up, the cars were carried all the way to the top of the vertical circles, where they would linger agonizingly before dashing madly back downwards. Eventually, the cars would begin to move in complete vertical circles in opposite directions. After many revolutions the process reversed and the motion died out."

Charles H. Freeman of Chandler, North Carolina, told me of the White Lightning ride at Carowinds Amusement Park near Charlotte. He described the accelerations involved in terms of g, the acceleration due to gravity. "White Lightning starts at ground level. The car is given a straight and level acceleration that is probably 2g or so for several seconds. When the acceleration ends there is a moment of panic as the body feels the sudden change of the force vector from near horizontal to normal vertical.

"Almost immediately the vertical (to the body) vector increases as the car enters a vertical loop. The value of the force vector decreases to slightly less than g at the top of the loop and increases again through the bottom. The car then enters a section of track that climbs at about 45 degrees until the car stops, by necessity, higher than the top of the loop. The car then travels backward down the ramp, through the loop, past the starting point, and, with braking, up another ramp to a braked stop, then slowly forward down to the starting point where it stops." Thus the passengers travel through a vertical loop once in the forward direction and then in the reverse direction. This is not a ride for me.

Freeman also told me of a ride called the Wild Mouse that neither of us has seen for years. "The ride looks like a small roller coaster, the highest altitude is about 15 feet, and the track fits in a square less than 50 feet on a side. The single car holds two good friends. The outstanding feature of the car is that the wheels are set well back from the front so that the passengers' feet are for-ward of the wheels. The car travels slowly, no more than 35 miles per hour, but the track makes very sharp turns. At each turn the car overhangs the corner so that the continuing track is almost di-rectly off the rider's shoulder before the car turns. Even the confident rider 'knows' that the car missed the turn.

"I classify rides using three parame-ters: falls and discontinuities, high accel-erations, and anticipations. Rides can also be rough or smooth. Falls are defined as times when the force of the seat on the rider approaches zero. Note that the seat force may never be zero for more than about a quarter of a second, or the riders would be thrown from the car and killed.

"Flat roller coasters are mostly rough, with many falls and anticipations. Loop-ing roller coasters are mostly smooth with high accelerations and a few falls. The heavy padded bars pressing you down are for effect to increase antici-pation. The Scrambler has an oscillating acceleration (sometimes high) with slight anticipation. The Rotor and the Enter-prise have smooth, high accelerations. White Lightning has a smooth, high ac-celeration with that startling discon-tinuity. The Muzek Express is rough with high accelerations.

"The swings you described and the fer-ris wheel are almost pure anticipations. Anticipation is generated when there is little in front of or under the rider. Speed also increases anticipation.

"The roller coaster at Walt Disney World in Florida, called Space Moun-tain, I believe, is rough with falls as usual. What makes this ride unique is that it is in the dark with no anticipations. Now, that's a ride."

2 Racquetball

Success in racquetball is enhanced by knowing the physics of the collision of ball with wall

A four-wall game such as racquetball, squash or handball demands of the player a great deal of skill in judging angles and bounces. The ball comes off the wall in a direction determined by the physics of the collision. An understanding of this physics enables a player to predict the ricochet of a ball approaching him and to calculate the ricochet he would like to achieve in order to put the ball out of the reach of his opponent. In discussing these phenomena I shall call to my aid some strange related tricks that can be demonstrated with a highly elastic solid ball sold in toy stores.

The toy ball is almost perfectly elastic: if you drop it, it bounces back nearly all the way to your hand. (A perfectly elastic ball would return to its initial height.) The ball also has a rough surface, so that when I throw it along the floor, it does not slip. Because of the ball's elasticity and roughness, it can be bounced in some surprising ways.

When I throw the ball downward at an angle, it bounces across the floor in a repeated pattern of high, short hops and low, long hops. If I put some spin on the ball as I throw it, it bounces to the left and right until it runs out of energy. The most startling demonstration involves throwing the ball under a table. A smooth ball would bounce between the table and the floor until it reached the far side of the table. A rough elastic ball bounces back to the thrower.

To study how a ball collides with a surface I first considered a uniformly solid ball bouncing on a floor. Suppose the ball approaches the floor moving to the right and downward. It helps to describe the velocity as being in two parts, one part parallel to the floor and the other part perpendicular. In addition the ball can be spinning about its center. A clockwise spin is a negative rotation and a counterclockwise spin is positive.

The ball's kinetic energy is in three parts, one part for each component of the velocity and one for the spin. If the ball is completely elastic, the collision does not change the total kinetic energy.

(The total kinetic energy is said to be conserved.) Only an ideal ball and collision follow this rule. In practice some kinetic energy is lost by being converted into other forms of energy. For example, some of it might end up in the vibrations of the ball. I shall ignore such losses and concentrate on the movements of a totally elastic ball.

The collision of the ball with the floor changes the perpendicular velocity in a simple way: it reverses the direction but leaves the magnitude and the associated kinetic energy unaltered. The parallel velocity and the spin are altered in more complicated ways. Still, the total kinetic energy is unchanged. An elastic collision might decrease the spin, but the parallel velocity would then be increased just enough to keep the total kinetic energy constant. This requirement of conserving the total kinetic energy is a strong tool for predicting the rebound.

Another important point is that the total angular momentum is conserved. One contribution to the angular momentum comes from the spin. This contribution is equal to the rate of spin multiplied by the ball's moment of inertia. The spin angular momentum is considered to be negative if the spin is clockwise and positive if it is counterclockwise. The moment of inertia depends on the mass of the ball and the way the mass is distributed. For a solid ball of uniform density the moment of inertia is two-fifths of the product of the mass and the square of the radius.

The other part of the angular momentum depends on how fast the ball is

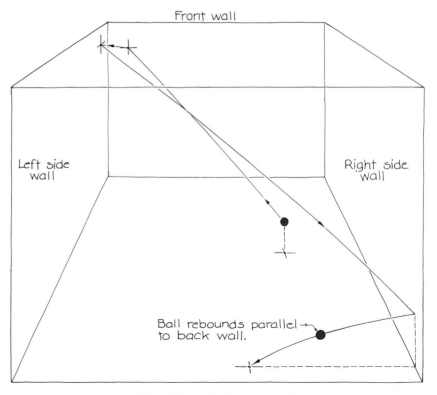

The troublesome Z shot in racquetball

moving parallel to the floor at the instant it touches the floor. This contribution to the angular momentum is equal to the product of the ball's mass, the parallel velocity and the radius. If the parallel velocity is toward the right, the contribution is negative; toward the left it is positive. The collision may change the two contributions to angular momentum in both magnitude and sign, but the total angular momentum remains. In sum, regardless of how the ball is thrown to the floor or how it spins, the total kinetic energy and the total angular momentum must remain constant in an ideally elastic collision.

The easiest demonstration is to drop the ball to the floor. If it has no spin initially, it must bounce back to your hand without spin because of the conservation rules. The only kinetic energy it has is associated with its perpendicular velocity. Since that velocity is only reversed by the collision, without any change in magnitude, the kinetic energy is unchanged. None of it can be transferred to the spin or to parallel velocity, and so the ball must travel straight upward. This result also satisfies the requirement that angular momentum be conserved. Before the collision and after it the ball's angular momentum is zero.

Suppose you put a clockwise spin on the ball. The collision directs the ball onto a new path. At the collision with the floor the spin creates a friction force toward the right, reversing the direction of spin. Because of the friction force, the ball also acquires a parallel velocity, so that it bounces to the right. The energy for the parallel velocity is taken from the energy of the initial spin.

Energy is also transferred when the ball is thrown to the floor at an angle and without spin. I had expected the path after such a bounce to be just as steep as the initial path, but it is steeper because the collision reduces the parallel velocity, converting some of its kinetic energy into spin energy. In terms of angular momentum the collision reduces the amount associated with the parallel velocity and increases (from zero) the amount associated with the spin. The total kinetic energy and the total angular momentum are conserved.

The steepness of the path after a collision depends on the initial steepness and the spin. When the initial spin is negative (clockwise), the final steepness is less than it is when the ball is thrown down without spin. A strong spin directs the ball along a low path over the floor. When the initial spin is positive (counterclockwise), the ball may bounce forward in a steep path, upward perpendicular to the floor or even backward, depending on the strength of the initial spin. The bounce is straight up if the ball initially has just the right amount of positive spin. (The product of the spin and the radius of the ball must be equal to

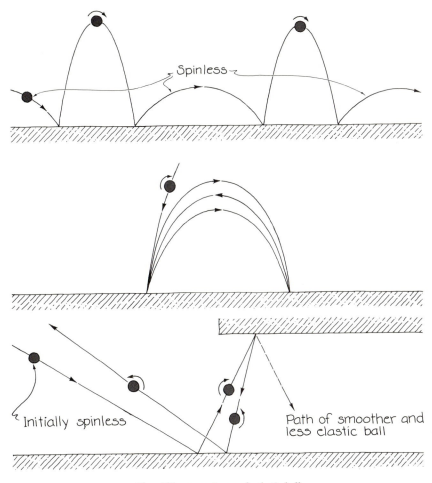

The odd bounces of a rough, elastic ball

three-fourths of the ball's initial parallel velocity.) With more counterclockwise spin the ball rebounds to the left. If the spin is less than the threshold amount, equal to zero or negative (clockwise), the rebound is to the right.

The steepness of the rebound can be understood in terms of the friction where the ball touches the floor. The friction force is opposite to the direction in which the surface of the ball is moving. At the moment of contact the surface motion has two sources: parallel velocity and spin. The friction opposes the sum of these two motions. For example, if the ball is thrown down at an angle and without spin, the surface touching the floor is moving to the right. The friction force acting on the surface is toward the left, which reduces the parallel velocity. The ball bounces toward the right with less rightward velocity than it had before the collision. Since the amount of perpendicular velocity is unaltered by the friction, the ball bounces in a path steeper than the one it followed in approaching the floor.

I also considered events in which the ball makes several bounces on the floor. Suppose the ball is thrown to the right without spin. The first bounce reverses the perpendicular velocity (so that the

ball goes upward), decreases the parallel velocity and imparts a clockwise spin. The ball rises to its maximum height and falls back to the floor. The surprising feature is that this second bounce restores the initial spin (which was zero) and parallel velocity. The result is the same regardless of the initial values of spin and parallel velocity. If the ball continues to bounce along the floor, its initial values of spin and parallel velocity are restored after every even number of bounces.

The phenomenon was readily apparent in the action of the elastic toy ball. I painted the equator of the ball so that I could monitor the spin. When I threw the ball to the floor with no initial spin, the first bounce was high and short, so that the ball did not move very far horizontally before the next bounce. The spin was clockwise. The second bounce was low and long. The ball had essentially no spin. Thereafter the ball repeated the pattern of a high, short bounce followed by a low, long one. Since the ball was not totally elastic, each bounce was less energetic than the preceding one. A perfectly elastic ball would periodically resume its initial spin of zero and its initial parallel velocity.

The interactions of spin and parallel

velocity account for the strange actions of a ball thrown to the floor so that it strikes the underside of a table. If the ball is initially without spin, it bounces from the floor on a steep path with a rapid clockwise spin; when it hits the table, it rebounds to the left with a counterclockwise spin. The second bounce from the floor is also to the left with a counterclockwise spin. The perpendicular velocity has been reversed three times but is unchanged in amount. The parallel velocity is now toward the left and is almost unchanged in amount. Hence the ball almost returns to the launch site.

Suppose the ball were smoother and less elastic. The first bounce would result in a weak spin and the second (from the underside of the table) would not be to the left. The ball would continue to travel to the right until it exhausted its kinetic energy.

I next turned my attention to an ideally elastic, hollow racquetball. Such a ball should perform all the tricks of a solid ball, although the spin values differ because the hollow ball has a different moment of inertia. If the ball is thrown at an angle to the floor (toward the right), it will hop straight up provided the spin is counterclockwise and the product of the spin and the ball's radius is equal to one-fourth of the parallel velocity rather than three-fourths.

In racquetball the serve comes off the front wall of the court. The ball rebounds to the opponent either directly or by bouncing from the side walls. The opponent must return the ball to the front wall before it bounces twice on the floor. Except on the serve, the ball can also be bounced from the back wall and the ceiling. I shall consider the shots that are allowed after the serve.

A player can impart spin to the ball with the racquet in only two ways: by stroking forward and over the top of the ball (achieving topspin) or forward and along the bottom of the ball (achieving backspin). The illustration at the left on page 11 depicts the spins from a view on the right side of the court.

Consider a ball hit hard and low toward the front wall with topspin. The collision is similar to one I described for a solid ball. The topspin (clockwise in the illustration) creates an upward friction force that directs the ball upward and reverses the spin. When the ball returns to the floor, the counterclockwise spin forces a low bounce toward the rear of the court. The potential advantage of such a shot is that your opponent may not expect the high rebound from the front wall or the low hop from the floor.

If you hit the ball hard and low toward the front wall with backspin, which is counterclockwise, it bounces toward the floor with a clockwise spin. It hits the floor close to the front wall and rebounds steeply upward. The potential advantage of this shot is that your opponent may not be able to reach the ball before it bounces from the floor a second time.

Usually my stroke gives the ball little or no spin, but it ends up spinning as soon as it bounces from a wall or from the ceiling. Consider a ceiling shot, which I often make to change the pace of the game. My opponent must adjust not only to the new path but also to strange hops off the floor. Suppose I make the ball bounce from the front wall to the ceiling. It leaves the ceiling with a clockwise spin. When it hits the floor, its parallel velocity is sharply reduced, making it bounce almost straight up. My opponent, who is expecting a rebound path resembling the path of the approach to the floor, waits too far back in the court.

If I make the ball bounce from the ceiling to the front wall, it approaches the floor with a counterclockwise spin. The collision with the floor increases the parallel velocity, sending the ball into a low hop. Again my opponent misjudges the rebound path and misses the ball. Both ceiling shots are better if I start them from about midcourt. Then the spin as the ball approaches the floor is strong and the strange hop is enhanced.

Suppose the ball is bounced off the front wall so that it moves toward the left side of the court. If you take an overhead view and ignore the curvature of the path due to gravity, the arrangement is similar to the one in which a solid ball is thrown at an angle to the floor. The collision reverses the perpendicular velocity (in this case the velocity perpendicular to the front wall), decreases the parallel velocity (the velocity toward the left side wall) and imparts a clockwise spin. In the overhead view the final path is steeper with respect to the front wall than the initial path because of the reduction in the parallel velocity. An opponent can quickly learn how to deal with this type of rebound in racquetball.

A more difficult shot to anticipate is one that bounces from two walls. Consider an overhead view of a shot in which the ball bounces from the front wall and then from the left side wall. The first bounce gives the ball a clockwise spin and a velocity directed toward the rear wall. Can you make the ball rebound from the side wall in any direction you choose or is the final angle of rebound fixed? Can the final spin be zero or any value of clockwise or counterclockwise rotation? To answer these questions I employed some mathematics published independently by Richard L. Garwin of Columbia University

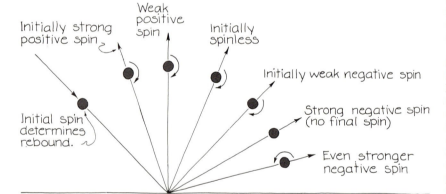

How the bounce depends on the initial spin

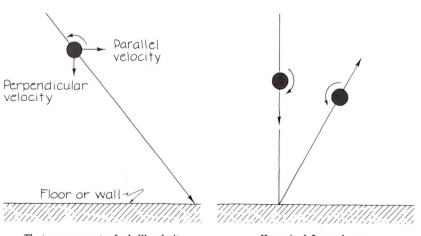

The two components of a ball's velocity   *How spin deflects a bounce*

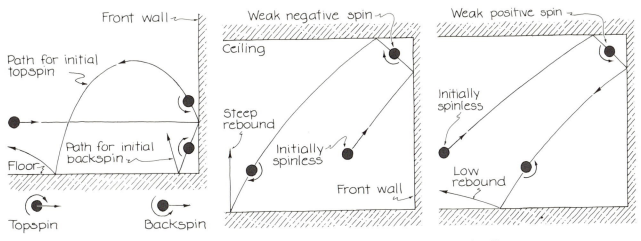

The energetics of topspin and backspin   *Schemes for making use of the ceiling*

and George L. Strobel of the University of Georgia.

Assume the ball is launched toward the front wall with no spin and has a small initial perpendicular velocity. You can make such a shot if you are near the front of the right side wall. Then an ideally elastic racquetball rebounds from the left side wall at an angle of about 12 degrees. If you are closer to the center of the court, the initial perpendicular velocity is larger and the angle of the rebound from the left side wall is smaller; the ball travels along the wall to the rear of the court.

You can use this arrangement to advantage. Suppose your opponent is near the middle of the right wall. By bouncing the ball off the front wall and into the left side wall so that it travels along the side wall to the back of the court, you can make it almost impossible for him to return the shot. Even if he is not far from the final path of the ball, the rebound off the side wall might at least prove confusing.

When I tested my calculations with a real racquetball, I found an approximate agreement. The steepest angle of rebound from the side wall was larger than the 12 degrees I had predicted. As I increased the initial perpendicular velocity by moving from the right wall toward center court, the angle of rebound decreased until the ball almost hugged the left wall on its way to the rear of the court.

The discrepancy between the actual and the predicted rebound off the side wall arises from the inelastic collisions of a real racquetball. If the ball hits a wall squarely, it compresses uniformly, storing its energy as elastic potential energy. Only part of the energy is reconverted into kinetic energy as the ball pushes off from the wall, again taking the shape of a sphere. A racquetball might bounce back with 60 percent of its energy in such a collision. The perpendicular velocity would then be about 80

percent of the initial value. (The change in velocity is proportional to the square root of the change in energy.)

A glancing collision is more difficult to interpret because the compression of the ball is not uniform and depends on the angle of the collision. The loss of kinetic energy and angular momentum reduces both the spin and the parallel velocity. (When the ball skims along the wall or the floor in an extreme glancing shot, you can hear the energy loss as a high-pitched squeal as the ball skips over the surface.) In my calculations I chose to reduce the spin and the parallel velocity after a collision by .4. With these reductions I found closer agreement between my predictions and the actual rebounds.

I was also able to explain why a real racquetball does not return to me when I throw it under a table. The reductions in energy and angular momentum in the bounces from the floor and the underside of the table trap the ball into bouncing almost vertically until it exhausts its kinetic energy.

Is there a way to hit the ball to the front wall so that it rebounds from a side wall parallel to the front wall? With such a shot you could win every game because your opponent could not possibly get to the ball in time. As it turns out such a shot is impossible. A rebound from a side wall is always toward the rear of the court.

Can a rebounding ball have any direction of spin or even no spin? Yes, because its final spin depends on the initial ratio of perpendicular and parallel velocities. For a perfectly elastic racquetball a spin of zero results when the ratio is 1 to 5. A smaller ratio yields a clockwise spin (from an overhead view), a larger ratio a counterclockwise spin.

The *Z* shot is a three-wall rebound that is marvelous to watch. When it was first introduced in the early 1970's, it confounded even the most experienced players. The ball is hit to the top left side

of the front wall, bounces to the left side wall, crosses the court to the rear of the right side wall and then rebounds parallel to the back wall. An opponent will need experience to anticipate the final rebound, but even then the ball will be difficult to return to the front wall. If I hit the *Z* shot less than perfectly, the ball might still be difficult to return if it hits the floor and then the back wall. My opponent must catch it near the back wall before the ball makes its second bounce on the floor.

Initially I thought a perfect *Z* shot was impossible. I doubted that the final rebound could be made to move parallel to the back wall. Armed with my mathematics I set out to follow the bounces.

I immediately met a problem. If the ball is assumed to be perfectly elastic, it rebounds from the left side wall at such a small angle that it hits the back wall instead of the right side wall. I factored an extra-long court into my calculation. I also ignored the curve resulting from gravity and made the calculation as though the ball remained in a plane parallel to the floor.

To launch the *Z* shot a player stands near the right wall at about midcourt. The ball is hit to the top left side of the front wall about three feet from the corner and three feet from the ceiling. Since such a shot makes the ball leave the left side wall with a clockwise spin, its collision with the right side wall creates a friction force toward the front wall.

Consider the velocity and the spin of the ball just before and just after the collision with the right side wall. The perpendicular velocity is reversed, directing the ball toward the opposite side wall. What happens to the spin and the parallel velocity? The collision is similar to one I considered earlier. The friction during the collision opposes both the spin and the parallel velocity, reducing the parallel velocity and reversing the spin. Under the proper conditions the parallel velocity can be reduced to zero, so that the ball's path is perpendicular to the side wall. This is how a perfectly executed Z shot makes the ball travel parallel to the back wall.

When my calculations include the loss of energy with each collision, my predictions are closer to the actual path of a Z shot in a court of the proper dimensions. The possibility of a final rebound parallel to the back wall is still present. My calculations are flawed, however, since the actual path has three dimensions. My assumption of a flat trajectory simplifies the calculations because the axis about which the ball spins is always kept parallel to the wall. In the actual flight of the ball the spin axis is often at an angle with respect to the side walls.

The around-the-walls shot also hits three walls. The ball is bounced from the right side wall to the front wall and then off the left side wall. The shot is designed to confuse an opponent, but if

the ball ends up at midcourt, he may have an easy chance of returning it to the front wall. I wondered if there was any way I could set up the around-the-walls shot to make the ball rebound from the left side wall parallel to the front wall. Expecting the ball to come to the rear of the court, my opponent would surely be caught off guard by this strange rebound.

I tried the shot in many ways without success. I wondered if the problem was my lack of playing skill, and so I turned again to mathematics. My calculations showed that such a rebound is possible if the ball begins with much energy and makes a small angle with the right side wall. If I had made the calculations earlier, I could have saved myself many futile swings of the racquet.

Many more shots can be studied with either a solid ball or a hollow racquetball. Perhaps there are some clever shots that even the professional racquetball players have yet to discover. You may be interested in studying how a ball loses energy in a glancing collision with a wall. You may also be interested in following the flight of a ball in three dimensions, so that the spin axis is no longer parallel to the walls. For this purpose a computer simulation of racquetball would be helpful. Be careful if you experiment with a solid, highly elastic ball in a racquetball court. I tried it just once. The ball moved and rebounded so fast that all I could do was get out of the way.

NOTES

Howard Brody of the University of Pennsylvania and Joseph P. Straley of the University of Kentucky have pointed out an embarrassing error in my article. I argue that in an elastic bounce of a ball on a floor the vertical velocity is reversed but unchanged in size. Thus the ball should bounce equally high on every subsequent bounce.

As obvious as this conclusion was to me, I forgot it when I observed the

bouncing of a highly elastic ball across my kitchen floor. When I put spin on the ball, it seemed to bounce in a repeated pattern of high and low bounces (as shown in the second figure), thus contradicting my previous conclusion. What I saw was an illusion. The heights of all bounces were actually about equal. However, the interaction of the ball's spin with the floor periodically altered the angle at which the ball rose. When the horizontal velocity was small after a bounce, the ball rose at a steep angle. When the horizontal velocity was large, the angle was shallow. I misinterpreted these angles, concluding that the height was small when the angle was shallow. The illusion was so captivating that I forgot my physics.

Brody also pointed out that in tennis the role of spin in the bounce of a ball is not as simple as I assumed in racquetball. "The high bounce of a top-spin shot (or low bounce of a back-spin shot) is not due to the spin interaction when the ball hits the court. It is caused by the vertical velocity component of the ball being higher (or lower) than a spinless shot with the same peak in its trajectory. That extra speed is due to the Magnus effect." This effect, due to the interaction of the spinning surface with the passing air, results in unequal air pressures on opposite sides of the ball. Depending on the spin of the ball, the pressure difference can give the ball lift or push it downward. If such an effect is important in the flight of a racquetball, then what trajectory the ball takes in the court is more complicated than I have assumed.

Brody has also recently analyzed how the surface of a tennis court determines the bounce of a ball. In particular, he studies what is meant by a "fast" and a "slow" court. You might like to investigate how his results affect a game of racquetball. His paper "That's How the Ball Bounces" is listed in the bibliography.

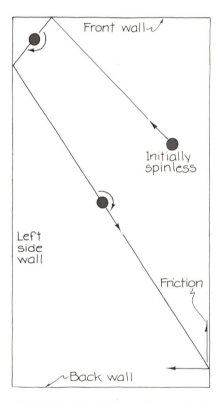

The Z shot as it would be seen from overhead

3

Billiards and Pool

The physics of the follow, the draw and the massé

Billiards and pool have the feel of physics. Balls collide with each other and the rails of the table like gas molecules in some two-dimensional tank. Actually the physics of billiards and pool is subtler. For example, a skilled player can impart spin to the ball, achieving such effects as the follow, the draw and the massé. Indeed, the interaction of the cue and the ball may be the most challenging application of classical mechanics. To master the forces and trajectories of both billiards and pool one must play the game often and analytically. A helpful step toward that objective is an understanding of the physics of the games.

Recently Todd King of Temple City, Calif., sent me his analysis of some of the classic shots in pool. Until a few years ago almost the only study concerning the dynamics of billiards was in lecture notes by Arnold Sommerfeld, who is better known for his work on early quantum mechanics. Last year David F. Griffing of Miami University devot-

ed a chapter to pool and billiards as part of his book *The Dynamics of Sports: Why That's the Way the Ball Bounces.* These three sources inform the following discussion of the physics of billiards and pool. After describing some simple relations I shall take up a few of the famous trick shots outlined in *Byrne's Treasury of Trick Shots in Pool and Billiards,* by Robert Byrne.

When the cue strikes the cue ball, both horizontal and rotational motions are imparted. For simplicity assume that the cue stick is horizontal and delivers only a horizontal force. Although the force can be applied anywhere on the ball's surface facing the player, the shock sets the ball in horizontal motion just as if the force were applied at the center of mass.

Now assume that the stroke by the cue is in the vertical plane passing through the ball's center of mass, namely on an imaginary vertical line running through the center of the face toward the player. The location of the blow along this line

has no direct bearing on either the ball's initial velocity or its momentum (the product of mass and velocity). They are set by two other factors in the collision. One factor, over which the player has virtually no control, is the duration of the collision. The second factor, easily controlled by the player, is the force on the ball. A "hard" shot generates more velocity and momentum than a "soft" shot because the force in the collision is greater.

In addition to horizontal motion the cue also generates a torque that makes a ball rotate about its center of mass. The magnitude of the torque is equal to the product of the force and a lever arm that represents the vertical distance between the middle of the ball and the point where the cue strikes. The torque increases as the distance of the blow from the middle of the ball increases.

The torque determines the initial rate at which the ball spins about the center of mass. The spin is proportional to the torque divided by the ball's moment of inertia (a number that takes into account not only the mass of an object but also the distribution of the mass around the axis of rotation). For a cue ball rotating about an axis through its center of mass the moment of inertia is two-fifths of the mass multiplied by the square of the radius. The factor of 2/5, which arises from the shape of the ball, plays a role in a player's decision about where to stroke the ball in certain shots.

If the player wants an initially nonspinning ball, he should strike it at the height of its center of mass. The lever arm for such a stroke is zero, and so the torque and spin are zero. With a higher blow the collision has a lever arm and hence a measurable torque. The ball moves forward because of the force of the collision, and it spins about its center of mass because of the torque. The ball has topspin: the top of the ball moves away from the player faster than it otherwise would. Striking the ball below the center results in backspin.

The player's stroke therefore controls

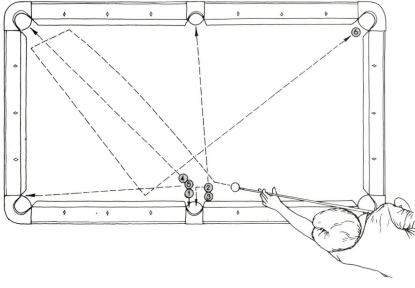

The "just showin' off" shot

three features of the motion. The force determines the velocity of the ball over the table. The lever arm of the force determines the direction of spin. The product of the force and the lever arm determines the rate of spin.

Without friction from the surface of the table the cue ball would continue moving until it hit a rail or another ball. Even a surface worn smooth from play can provide significant friction, however, if the ball slips on the cloth. The friction can be high enough to alter both the horizontal and the rotational motion of the ball and thereby significantly change the shot. If the ball rolls over the table without slipping, the friction is low and affects little more than the maximum distance of roll.

The friction on a slipping ball depends on the weight of the ball and on the surface texture of the cloth and the ball, but it is independent of the rate at which the ball slips. The direction of the force depends entirely on the direction of slip. Suppose the player delivers a large topspin to the cue ball; the bottom surface then slides toward him and the center of mass moves away. At the point of contact the friction force is opposite to the sliding. (The force is away from the player.) Since the friction opposes the sliding, it begins to decrease the spin of the ball about its center of mass. And since the friction force is away from the player, it continues to propel the ball forward and away from the player. A ball given a large topspin can run for a long time because of this additional propulsion.

Suppose the player imparts backspin. The force sends the ball away and the torque sets it into a spin that makes the bottom slide over the cloth in the same direction. The friction force on the bottom on the ball is thus toward the player. Again the friction tends to slow the spin, but this time its effect on the center of mass is rearward. As the ball slides, its forward motion and spin are slowed. Eventually the backspin is eliminated and the ball begins to roll without sliding. A ball hit with backspin runs only a short distance because the friction opposes the motion of the center of mass.

A ball with topspin will slip unless the speed of its center of mass is equal to the spin multiplied by the radius. Then the ball's forward motion is exactly matched by the motion of the bottom surface through the point of contact with the table. To achieve this match instantly the player must stroke the ball at a point that is exactly above the middle by a distance equal to two-fifths of the radius. This relation is established by the fact that the 2/5 in the formula for the moment of inertia must be canceled by the 2/5 in the formula for the lever arm of the torque.

If the cue ball is struck higher than at this special point, the spin rate is at first

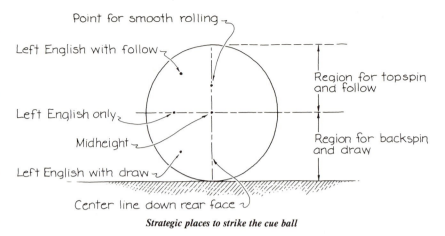
Strategic places to strike the cue ball

too large for a match to be achieved, but friction tends to force the ball into the matched state. The friction (directed forward) reduces the spin and increases the speed of the center of mass. If the ball does not run into anything, the match of speeds is eventually made and the ball begins to roll smoothly.

If the ball is struck between the middle and the special point, its spin is in the right direction for smooth rolling but is too small. At the point of contact with the cloth the ball's surface has a small rearward motion owing to the spin and a larger forward motion owing to the speed of the center of mass. The net slipping motion there is forward, creating a rearward friction force that tends to increase the spin and decrease the motion of the center of mass until the match that causes a smooth roll is achieved.

If the ball is struck below the middle, the spin is in the wrong direction for smooth rolling. This time the friction induced by slipping reduces both the spin and the speed of the center of mass. Eventually the spin stops and the ball begins to roll smoothly.

A skilled player can impart a long or short run to a ball by striking it at a point relative to the special point for smooth rolling. If he wants the ball to reach the far side of the table quickly, he must strike it above the special point so that the friction propels the ball.

The player is more likely to be concerned with how the ball is rotating when it strikes another ball. (The other balls are called the object balls.) A collision between a cue ball and an object ball transfers momentum from the cue ball. In a head-on collision the transfer is complete, leaving the cue ball with a motionless center of mass. In a glancing collision the cue ball loses only part of its momentum and continues to travel. In any collision virtually none of the cue ball's rotation is transferred because the friction between the surfaces of the colliding balls is minute and the collision is brief. Only with significant friction could the cue ball transfer spin to an object ball.

The absence of transferred spin leads to two interesting shots. Suppose the cue ball is hit with topspin and collides head on with an object ball while it is still sliding. Just after the collision the cue ball's center of mass is motionless but the ball continues to spin. The forward-directed friction generated by the spin slows the spin and begins to propel the center of mass. Soon the cue ball begins to roll again, following the object ball. This is a follow shot. A cue ball with topspin is often said to have "follow" or "follow English."

If the cue ball is given backspin, it will return to the player after hitting an object ball head on. The collision leaves the cue ball with a motionless center of mass but with the same amount of spin. The friction generated by the sliding surface is toward the player. As the friction slows the spin and propels the center of mass, the ball begins to roll smoothly toward the player. This is a draw shot. A ball with backspin is often said to have "draw" or "draw English."

A follow shot is depicted in the middle illustration on the next page. A player who wants to pocket the four ball and the seven ball with a single shot strikes the cue ball with follow, thus causing the four ball to ricochet off the seven ball and into the pocket. The collision of the cue ball with the four ball leaves the cue ball momentarily spinning in place, but sliding friction soon propels it toward the pocket again. In the meantime the seven ball has

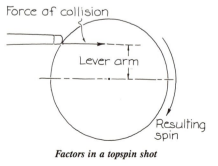
Factors in a topspin shot

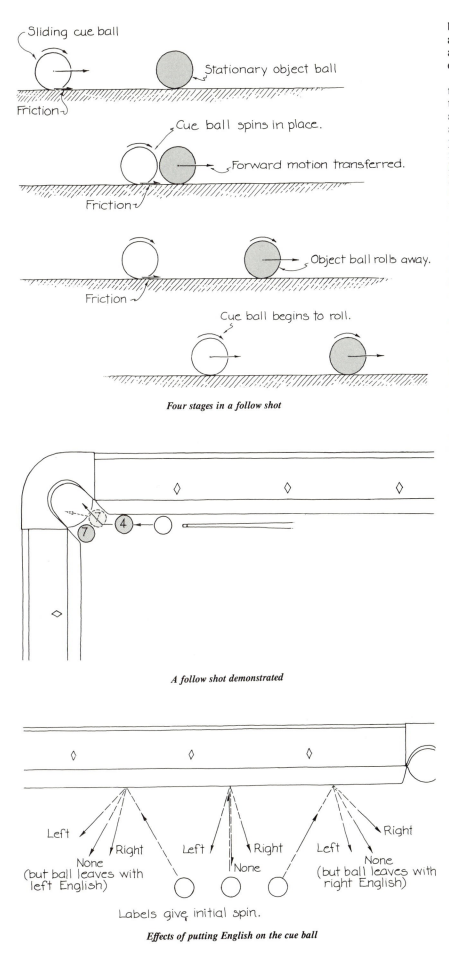

Four stages in a follow shot

A follow shot demonstrated

Labels give initial spin.

Effects of putting English on the cue ball

bounced off the rail near the pocket and come into line between the cue ball and the pocket. The cue ball then pockets the seven ball and comes to a stop.

So far I have written only of stroking the cue ball along a vertical line through the center. The result is topspin or backspin about a horizontal axis. If the ball is stroked elsewhere, the axis of spin still passes through the center of mass but is no longer horizontal. A stroke on the left side of the ball is said to give left English, on the right side right English. From overhead left English is a clockwise spin about the vertical. As before, the rate of spin depends on the lever arm associated with the force. The farther off center the ball is struck, the larger the lever arm is and the faster the spin around the vertical is. The friction of the ball with the table serves only to diminish the spin.

If the cue strikes the ball high or low on the side, the ball spins about an axis that is between the horizontal and the vertical. A stroke below the middle and to the left of center results in a draw with left English. The center of mass is given momentum in the usual way and the ball spins about an axis tilted out of the vertical toward the player's left. One can view this rotation as being two simultaneous spins, one that is clockwise about the vertical and another that is backspin about the horizontal. The primary friction force on the ball is imparted by the backspin.

A cue ball with side English travels in a straight line like a ball with no English, but the angle at which it rebounds from a rail is remarkably different. When the ball has no initial English, its angle of rebound is the same as its angle of approach to the rail. If the approach to the rail is perpendicular, as is shown in the bottom illustration at the left, the ball must retrace its path after rebounding. If it has left English, however, it returns along a path on the player's left because of friction during contact with the rail. From an overhead view the ball with left English turns clockwise. When it slides along the rail, the friction force is to the left. When the ball rebounds, it has not only its velocity perpendicular to the rail but also this leftward component. The ball returns along a straight path resulting from the combination of the two motions.

A left-English ball approaching the rail at any other angle is similarly redirected. The opposite effects arise with a right-English ball. A simple way to remember the difference is to associate the direction of English with the rotation of the rebound path: left English results in a clockwise rotation of the path and right English results in a counterclockwise rotation.

If a ball approaches the rail with no English and at some angle other than 90 degrees, friction from the rail gives it

English. Consider the ball's approach to be a combination of one motion perpendicular to the rail and another parallel to it. Once contact is made the parallel component generates friction on the ball. The resulting torque spins the ball about the vertical, imparting English. Suppose a ball is sent into the rail from the player's right. If it initially had no spin, it leaves the rail with left English.

All these rotations of spin are about an axis in a plane perpendicular to the direction of travel of the ball. The massé shot supplies spin about an axis out of that plane. With the cue almost vertical the player strikes downward on the side of the cue ball. The horizontal part of the stroke determines the initial path of the ball, but the spin given to the ball generates a friction from the table that ends up curving the path.

Suppose the player strikes sharply the left side of the ball. Since the blow is hard and the lever arm is large, the spin imparted to the ball is large. The spin is about an axis that is approximately in a horizontal plane but is not perpendicular to the initial path of the ball. To simplify the spin one can regard it as consisting of two simultaneous spins about different axes. One is parallel to the initial path and the other is perpendicular to it. The spin about the axis perpendicular to the initial path is similar to that of a simple draw shot. The spin about the other axis forces the ball to slip perpendicular to the initial path, thereby generating a friction force that is also perpendicular to the path. This sideways force curves the path of the ball.

The massé is commonly employed to send the cue ball around an obstacle to reach a hidden ball. A more complex massé shot is shown in the lower illustration at the right. The idea is to sink the 15 ball and the eight ball with a single shot and to have the eight ball enter the pocket last. The cue ball is given a massé shot, knocks in the 15 ball, misses the eight ball and heads toward the rail in a curved path because of sideways friction. After rebounding from the rail the ball stops and then reverses its horizontal motion, heading back to pocket the eight ball.

The initial stroking of the cue ball gives it both backspin for a friction force like the one in the standard draw shot and sideways spin for a sideways friction to curve the path into the rail. The rebound off the rail is little affected by the backspin, but the sideways friction keeps the ball near the rail. After the rebound the backspin finally stops the horizontal motion of the ball's center of mass. Since the ball still spins, that friction force then brings it back toward the player. The remaining sideways friction continues to drive the ball toward the rail. Hence after the reversal of path resulting from the drawlike component of the shot the ball comes back near

the rail to knock the eight ball into the corner pocket.

When a cue ball collides with an object ball, part of the momentum and kinetic energy of the center of mass of the cue ball is transferred. The transfers are almost total if the collision is head on. In a glancing collision the transfers cause the two balls to separate along approximately perpendicular paths. (In practice a small amount of energy is lost by the balls in the collision, and the angle between their paths is a bit less than 90 degrees. I shall disregard this complication.)

You can easily predict where the cue ball and the object ball will travel after a collision. Imagine the instant the two balls touch and mentally draw a line between their centers. At that instant the object ball acquires two forces from the cue ball. At the contact point and perpendicular to the line between the centers is a small friction force. It is almost always small enough to be ignored. Parallel to the line is a larger force that pushes the object ball off along a path

that is also parallel to the line. The direction given to the object ball depends almost entirely on the orientation of the line between the centers of the balls at the instant of contact. Through experience the skilled player can direct the cue ball so that it makes contact in just the way necessary to send an object ball into a pocket. The player can be certain the cue ball will travel perpendicularly to that path.

If the cue ball is given follow or draw and still has the associated sliding when it reaches the object ball, the collision changes somewhat in that the cue ball leaves the collision site on a curved path. Suppose the cue ball has been given a large follow. The collision transfers part of the kinetic energy and momentum of the center of mass. If one can disregard entirely the friction between the colliding balls, none of the spin of the cue ball is transferred to the object ball. The cue ball begins to move away from the collision site along a path perpendicular to the path taken by the object ball. The cue ball still has topspin. The curious

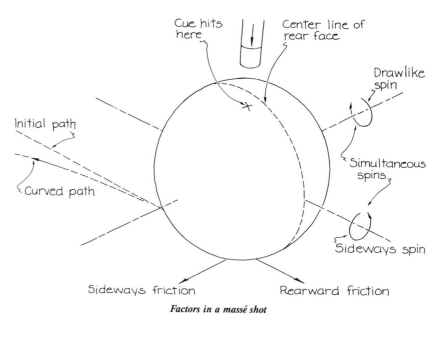

Factors in a massé shot

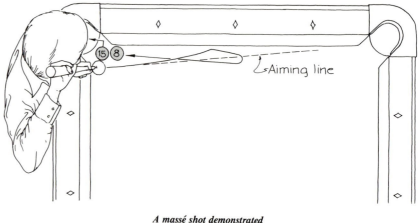

A massé shot demonstrated

feature is that the ball no longer slides parallel to its path. The component of the sliding perpendicular to the path provides a sideways friction force that pushes the ball into a curved path. Therefore when the cue ball is given follow, it tends to curve back toward its original direction after a collision. A cue ball with draw tends to curve away from the original direction.

Normally the friction between two colliding balls is negligible. It can be greatly increased if the surfaces are covered with chalk. My favorite example comes from Byrne's book. The illustration below depicts the setup: the player must get the five ball into the pocket at the right. Can the shot be made without contact between the cue ball and the spotted ball? Normally the shot is impossible. The five ball can travel to the pocket only if the player has aligned the collision so that the line joining the centers of the cue ball and the five ball points to the pocket. The spotted ball is clearly in the way.

The shot can be made if the player chalks the left side of the five ball and then sends the cue ball into the side with a little right English. With chalk on the collision area the friction between the balls is no longer negligible. The five ball is subjected to two forces during the collision, one force parallel to the

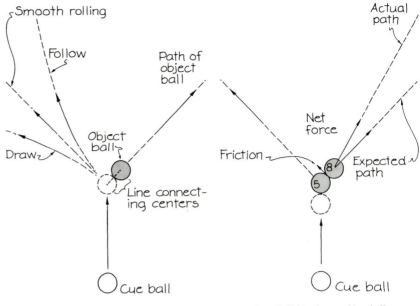

A cue ball hits an object ball   *A cue ball hits frozen object balls*

line connecting the centers and the other force (friction) perpendicular to that line. The five ball heads off in the direction of the net force, which by design is toward the corner pocket.

Chalking a ball is certain to get you thrown out of a pool game, but a similar application of friction between balls can

be achieved more acceptably. When a cue ball strikes an object ball that is already touching another object ball (the two object balls are said to be frozen), the collision creates a friction between the object balls that can significantly alter the path of one of them. Consider the situation in the illustration at the top right on this page. The cue ball is sent directly into the five ball, which is frozen to the eight ball.

It is best to analyze the collision in two steps. First the cue ball transfers energy and momentum to the five ball, which then collides with the eight ball. The five and the eight should separate along perpendicular paths, but the eight ends up traveling more in the forward direction because of friction between the object balls.

During the collision between the five ball and the eight ball the five begins to move perpendicular to the line between the centers. The eight ball should move parallel to that line because of the force of collision from the five. Since the balls are initially frozen, however, the motion of the five ball rubs the surface of the eight ball, generating a friction force that briefly pushes the eight ball perpendicular to the anticipated path. The actual path is then set by the combination of these two forces on the eight ball during the collision; the path is more in the forward direction than it would be if the balls were not initially frozen.

My last example, the "just showin' off" shot, has become famous because Steve Mizerak, a master of pool, performed it in a television commercial. Five balls are clustered around a side pocket. The six ball lies at the mouth of a corner pocket. Can all six of the balls be pocketed with a single shot? I certainly cannot make the shot, but

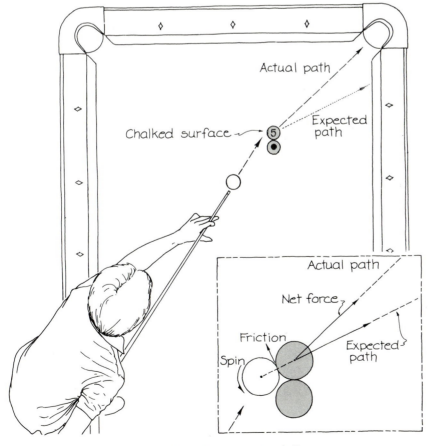

Creating friction between two object balls

Mizerak is said to be successful three times out of four.

The cue ball is sent into the two ball with follow and left English. Imagine the position of the balls and the forces between them at the instant the cue ball reaches the two ball. The two ball has three forces on it. One force is parallel to the line connecting its center with the center of the cue ball. Another is along the line connecting the centers of the two ball and the three ball. Because those balls were initially frozen, the two ball also has a rightward friction force on it at the point of contact when it slides off the three ball. The net force sends the two ball over to the five ball, where it ricochets into the side pocket.

Meanwhile the three ball has been moving. When the cue ball hit the two ball, the three ball received two forces from the two ball. A force parallel to the line connecting their centers knocked the three ball hard into the rail. The second force was leftward friction generated by the rubbing of the balls when the two ball departed toward the left. (This friction arises because the two ball and the three ball were initially frozen.) The rail pushes back on the three ball, directing it straight across the table. The ball also travels somewhat to the left because of its brief friction with the two ball. The three ball ends up in the pocket on the far side.

The five ball was initially frozen to the one ball and the four ball. When the two ball hits it, the five ball is subjected to several forces. One force is along the line connecting its center and the center of the two ball. Two more forces lie along lines connecting the center of the five ball with the centers of the one and the four. In addition the five ball has friction forces from being frozen to the other balls. The net force on the five ball is neatly toward the pocket in the left corner. The net force on the one ball is toward the nearby side pocket. The net force on the four ball is toward the far left pocket. Five balls are down.

After this cluster of balls has spread the cue ball returns to the area. It had been launched with follow (for a long run) and with left English. Its collision with the two ball left it traveling toward the far rail with most of its initial spin. The spin, however, is now somewhat sideways to the path. The cue ball curves to the left, rebounding closer to the corner pocket than it would without the sideways force. Its bounce from the rail removes most of the spin. Thereafter it travels in straight lines, bouncing twice more from rails. It finally reaches the six ball at the other end of the table and pockets it.

I have of course been describing only a limited selection of shots. Thousands of interesting shots remain to be ana-lyzed. You might be particularly interested in figuring out the physics of shots into large clusters of frozen balls. Byrne has several curious examples devised by 19th-century masters of pool and billiards. You might also be interested in jump shots, in which the cue ball is sent hopping over the table or even between two tables. Be careful. Proprietors of pool halls rarely tolerate such shenani-gans, even in the interest of scientific investigation.

NOTES

No one has conducted a detailed analysis of pool shots, especially on the matter of how spin influences the rebound of a ball from a rail. I have analyzed the details of a racquetball rebound in the preceding article. You might like to apply those results to pool. Still, the analysis will be theoretical. Do the predictions apply to a real situation in which the ball undergoes energy losses? To follow the motion of pool balls you might arrange to take stro-boscopic photographs of a pool table on which various shots are staged.

Martial Arts

4

*In judo and aikido application
of the physics of forces makes
the weak equal to the strong*

Judo and aikido are martial arts that demand an intuitive understanding of the physics of forces, torques, stability and rotational motion. This month I shall examine a few of the basic throws of these two forms of combat. Although I cannot convey fully the grace each throw requires, I can break it up into components that can be examined in terms of classical physics. The experiments I shall describe call for actual performance of the throws, but you should do them only under expert supervision. Both types of combat can be dangerous to you and your opponent.

In judo the main goal is to overcome your opponent's stability. The skill lies in the anticipation of his movements and the timing of your responses. The idea is to avoid forcing your opponent into a firm resistance to your throw that would pit your strength against his. A small but skilled judo player has a distinct advantage over a larger but unskilled opponent if a contest of strength is avoided.

Probably the best example of this advantage is the basic hip throw, which is most effective against a taller and slower opponent. In the normal judo competition you face your opponent with your hands grasping the lapels or shoulders of his uniform. To execute the throw you step forward with your right foot to a point between his feet, pulling him downward and toward your right. The throw works well if you have caught your opponent just as he has stepped forward with his right foot. He is still stable against a pull directly toward you, but he is considerably less stable against a pull to your right because of the position of his feet.

During your step forward you curve your body forward so that your head is at your opponent's shoulder level. Next you rapidly turn your left hip backward while pulling him onto your right hip. This should be the first body contact during the movement. If you continue the pull with your hands and the rearward turn with your left hip until you are facing in the same direction as your opponent, he will be rotated over your right hip and onto the mat.

Since you do not want to hurt your opponent in the sport, you maintain your grip on his uniform during the fall so that he lands on his left side and can slap the mat with his left arm during impact. The slap spreads the impact force over a larger area so that the stress on his ribs is not enough to hurt him. Part of the early training in judo involves timing the slap on the mat to coincide with the impact. The only time I have been hurt in the sport is when I failed to slap properly.

Timing and smooth execution are essential to the hip throw, but an understanding of the physics, particularly of the torques and the center of mass, is also necessary. Your opponent's center of mass is the geometric center of his mass distribution. It can be regarded as the point where the gravitational force acts on the body as a whole, which is why it is sometimes called the center of gravity. Your opponent is stable as long as his center of mass remains over the support area outlined by his feet. When he stands upright in a normal posture, his center of mass is approximately between his spine and his navel. Therefore he is stable until you force or trick him into moving his center of mass or into losing part of his support area.

Suppose during a throw you manage to move your opponent's center of mass forward of his feet. Even if you then no longer aid the throw, the gravitational pull on his center of mass creates a torque that might make him fall. To calculate a torque one must multiply two items: the force acting to bring about a rotation and the lever arm between the pivot point and the force. The lever arm is the perpendicular line from the pivot point to a straight line through the force. If you have made your opponent unstable, the force that can rotate him about his feet to the mat is the gravitational pull on him. I represent this force, which is merely his weight, by a vector pointing straight down from his center of mass. Here the lever arm is the horizon-tal distance between the pivot point at his feet and an extension of a vertical line running through the weight vector. The torque on an unstable opponent is the product of his weight and the lever arm. When your opponent is upright, the lever arm for his weight vector is zero and so the torque is zero too. When he is caught with his center of mass forward of his feet, the lever arm is no longer zero and the resulting torque causes his rotation. The longer the lever arm (the more he is leaning), the greater the torque. One of the objectives of judo is to trick your opponent into an unstable position so quickly that recovery is impossible. Once he is unstable you can continue the throw by applying another torque to him, one that will bring him to the mat long before he can even attempt to regain his stability.

During the hip throw you initially pull on your opponent's uniform to make him unstable. If you pulled directly toward you, you could not easily cause this instability because his center of mass would be moved over his forward foot. He could then maintain his balance by bending that knee. To make him unstable you would have to move his center of mass a relatively great distance until it was beyond his forward foot, but the motion would require a strong and prolonged pull, which he could counter quickly.

An easier way to make your opponent unstable is for you to pull to your right, because in that direction his center of mass must be displaced only a relatively short distance before it no longer lies over the support area. He will probably not be able to counter such a pull before the instability is established. Thereafter he will not be able to counter the continued rotation involved in the hip throw.

Your pull has an additional purpose. It curves your opponent's body so that his center of mass is brought forward to his navel or just outside his body. This new position will aid you in rotating his body over your right hip. Once body contact is made a new pivot point is established at your hip and your pull cre-

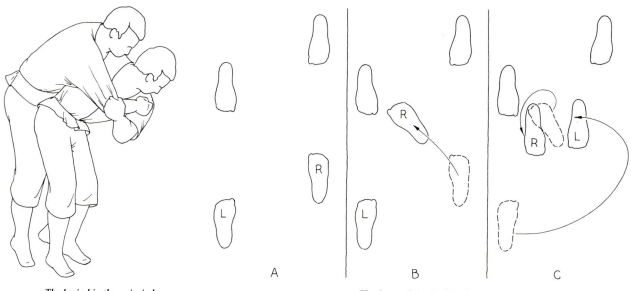

The basic hip throw in judo

The footwork in the hip throw

ates a new torque on the opponent, one that will cause the rotation over your hip. As before the torque is calculated by multiplying the force on the opponent by the lever arm between the line of that force and the pivot point. This time the force is your pull and the pivot point is your hip. Thus the hip throw gives rise to two torques on your opponent, one torque due to his own weight and unstable position and one due directly to the pull you are exerting on him. The throw begins with the first torque so that you can set up the second one without resistance from him.

Suppose you do not curve your opponent's body forward and bring his center of mass out to his navel. Then when you attempt to rotate his body over your

right hip, a torque due to his weight will actually counter the torque from your pull on his uniform. Suppose he is still in an upright position when you make body contact and attempt to rotate him over your hip. The pivot point for the rotation is your hip; I shall apply it in determining the lever arms for both of the torques then acting on the opponent.

One of the torques is the product of your pull and the lever arm from your hip to the line through the vector of the pull. The other torque is the product of your opponent's weight and the lever arm from your hip to a vertical line passing through the opponent's center of mass. If your opponent is standing upright, the torque from his weight opposes the torque you are applying,

since it attempts to rotate him in the opposite direction over your hip. To finish the throw you now must overcome the torque due to his weight, but the time required destroys your advantage in the surprise of the throw. Moreover, you must pit your strength against his.

When the hip throw is properly executed, you bring your opponent's body forward in a curve, move his center of mass out to his navel and so decrease or eliminate the lever arm associated with his weight. The torque due to his weight is therefore diminished and you have a comparatively easy time rotating his body over your right hip. The throw works better on an opponent who is taller than you are because you can pull him downward into the proper curved

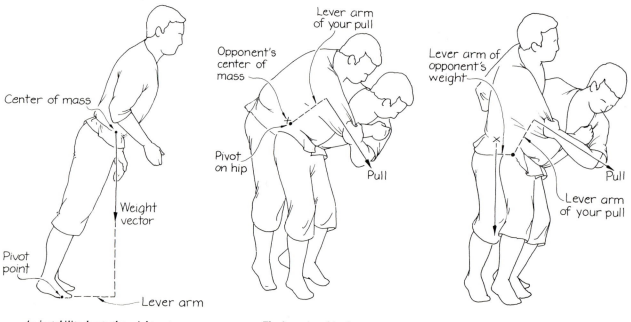

An instability due to the weight vector

The forces in a hip throw

An improper hip throw

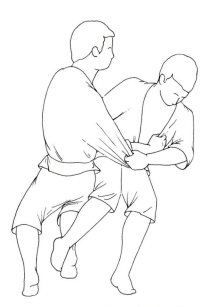

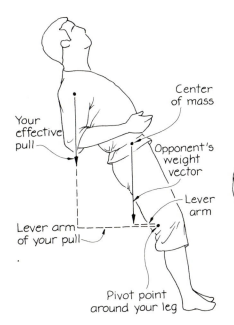

Center
of mass

Your
effective
pull

Opponent's
weight
vector

Lever
arm

Lever arm
of your pull

Pivot point
around your leg

The "major outer reaping throw"

The forces in the throw

An ankle sweep

posture more easily than you could an opponent who is your height or shorter. You can also more easily slide your right hip under a taller opponent. The lever arm of your pull on the uniform of a taller opponent will also be larger, thereby providing more torque to bring him over your hip.

The "major outer reaping throw" (it is called *osotogari* in Japanese) is somewhat easier to understand in terms of the rotational motions. As your opponent steps backward with his left foot you step with your left foot just to the outside of his right foot and pull downward on his uniform to force his weight downward on that foot. Your pull will also be toward his right rear so that his body is curved backward. He is already in an unstable position because your pull moves his center of mass to his right rear and away from the support area of his feet. He cannot escape by sliding his feet to the rear and regaining his balance because you have forced him downward. His instability results from the torque his weight creates around a pivot point at his feet, primarily his right foot. The lever arm runs between the pivot point and a line through his weight vector. Placing him in this position sets him up for the next part of the throw, in which you further remove his support area and apply a second torque to bring him rapidly to the mat.

You continue the throw by stepping with your right foot around and behind your opponent's right foot. Then you sweep your right hip and leg to your rear while you force his right side downward and to his rear, leaving him with virtually no support base. The two torques cause him to rotate about a pivot point on your right leg. Even if you did not continue to pull on him after sweep-

ing his leg, he would rotate about your leg because his center of mass is being pulled down by gravity. Your downward pull provides another torque to hasten his fall. In this throw the two torques complement each other.

The "sweeping ankle throw" (*okuriashi bari*) removes your opponent's leg support in a similar manner. As he is about to place his weight on his right foot in the course of a step forward or backward you sweep your left foot into that leg just above the ankle. Simultaneously you pull his uniform in his original direction of travel. Suppose he was moving forward. You pull forward (and so meet no resistance from him) as you sweep his right foot into his left one. Even if he manages to keep his left foot on the mat, his support area is greatly reduced and is swept from under his center of mass. His weight vector through the center of mass provides a torque that will take him to the mat. If you lower your body while maintaining your pull on his uniform, you will provide another torque that will rotate him to the floor. The pivot point is his left

foot, and again the two torques complement each other.

Advanced judo classes teach methods of disabling an opponent on the mat. Most of these "hold downs" entail trapping your opponent on the mat with your weight positioned in such a way that he cannot roll over or rise even if he is stronger than you are. For example, in the "cross armlock" (*udehishigi-jujigatame*) you are positioned with part of your weight on the upper torso of your opponent on the mat. He not only is prevented from rising but also probably will not even move for fear of having his arm broken.

The maneuver originates when you are astride your opponent, who is on his back. When he raises his left arm to ward you off, you grasp his wrist with both of your hands, fall to his left side, throw your right leg across his neck and with your left knee raised drive your left ankle into his side. His left arm is pinned between your legs with the elbow downward. Even a gentle downward push on his wrist creates a tremendous torque on his arm around the pivot point where

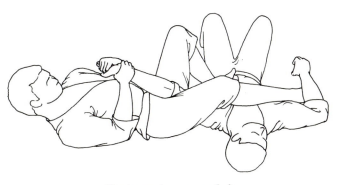

How to execute a cross armlock

that arm crosses your right leg. He cannot sit up because your weight creates an overwhelming torque on him as he attempts to rotate his trunk about a pivot point at his hips. He also cannot free his left arm even if he is considerably stronger than you are. He could try to counter your torque on his arm by using his shoulder muscles, but they would pull on his arm at approximately the location of the pivot point and so their pull would have a short lever. As in most judo techniques, a person trained in creating the correct torques on an opponent has a tremendous advantage even if the opponent is much stronger.

Aikido is a relatively modern form of martial art that incorporates techniques from many of the other martial arts. It is distinguished by its firm code of avoiding injury to the opponent. Hence it is a form of self-defense rather than a sport. It involves no techniques that can be regarded as attacks. I think it is the most difficult of all the martial arts to learn. Its demands for skill, grace and timing rival those of classical ballet.

Aikido employs many of the same principles of physics that are found in judo. Suppose your opponent grasps your wrists from behind. In one of the aikido maneuvers you smoothly lower your body while bringing your wrists

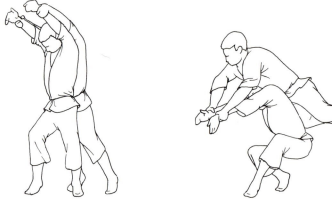

The aikido movement for escaping a hold on the wrists from behind

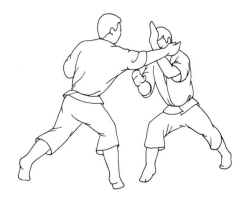

A defense against a hand slash

The defense against a hold on the arms from behind

upward and over your head toward the front. Your opponent hangs on to your wrists but is brought forward by your descent and slight lean forward. His position is therefore unstable because his center of mass is now slightly forward of his feet. You rapidly draw your right leg backward and drop onto your right knee. Your arms and torso are brought forward and downward in a large arc. Because the first part of the motion in-

duced your opponent to hold tightly to your wrists, he is now thrown over your body in a front somersault.

As in much of aikido, your opponent actually throws himself. He cannot prevent your forward motion because of the unstable posture in which you initially place him. Even if he has superior body weight, he cannot stop the motion by pulling downward on your raised wrists. In such a position he can pull only along the length of your arms. The torque due to such a pull is zero because there is no lever. Remember, the lever is the perpendicular from the pivot point (in this case your shoulder) to the line through the force. The line through your opponent's pull passes through your shoulder and therefore has no lever arm. Even if he is heavy or strong, he cannot rotate your arms once you have them properly over your head.

Many of the techniques of aikido employ the deflection of a force directed at you. Suppose your attacker throws a punch at your face. To stop the punch directly requires a large impact force, probably higher than 3,000 newtons. With such forces bones are likely to be broken. A wiser technique is to deflect the strike. Although a large force is necessary to stop the punch directly, only a small force is needed to deflect it. A force of 10 newtons may be enough to deflect a punch by a centimeter.

Although most followers of Western styles of fighting consider an attack to be an advantage, in aikido the attacker is at a distinct disadvantage because of the momentum of his strike. You can use his momentum to throw him to the mat. Suppose your attacker steps forward with his right foot and slashes at your face with the side of his right hand (a typical attack both in Western styles of fighting and in karate). You slide your left foot to the rear as you parry his slash with your left arm. The parry is meant to deflect the slash, not to stop it or even to slow it, since either effect would require strength from you. During the parry you guide your attacker's right arm downward into the grasp of your right hand. While still not fighting the forward momentum of his slash you pull him around in the circular motion you have begun with the withdrawal of your left foot. He was relatively stable against a pull directly forward because of his extended right foot, but he is highly unstable against a pull forward and to his left. In such a direction his center of mass does not have to be moved far before he becomes unstable against a fall. Therefore as you continue to circle you pull him in that direction. He now has two serious disadvantages. First, he is committed to a forward motion that would take a considerable force to stop, even from himself. Second, your pull and his motion are removing his center of mass from his base of support.

To complete the throw rotate your attacker's right arm downward while stepping to your left rear. Turn his wrist upside down and bend his hand around it. At this point it is impossible for him to prevent the throw. He is now off balance and completely unable to stop his own motion. He also cannot pull out of your grip because you have bent his arm at the wrist. Although his arms may be strong, he cannot prevent the torque you create when you push his hand around his own wrist. You bring him to the mat.

How would such a strike to the head be handled in karate? In the Korean karate style of *tae-kwon-do* I was taught to parry a slash with a powerful strike across the opponent's arm. Deflection was important but so was countering the slash with a large force. Force was working against force, and usually the stronger person won. (I was rarely that person.)

Circular motion is employed in aikido both for deflection and to aid in throwing an opponent off balance. Suppose someone approaches you from behind, reaches around your body and pins your arms to your sides. You should reach upward and hold his hands tightly to your chest while sliding your foot forward. The timing is critical because you want to move your torso forward at a rate matching the speed of your opponent. If you delay, you will lose the advantage of exploiting his momentum. If you move too fast, you will have to drag him forward. You must slide your right foot forward at the correct speed and then suddenly lean forward and rotate your body to your right.

The combination of your opponent's initial momentum and your rapid rotation throws him off to the right. He cannot prevent your throw because your lean forward brings his center of mass forward of his feet. He cannot release himself from the forward motion because of his established momentum, because you have pinned his hands and because of his grip around your arms. The centrifugal force on him during your rapid rotation is too large for him to deal with in his unstable position. Hence he essentially throws himself to the floor.

Two more examples of how aikido employs a small force to bring an attacker off balance entail stick fighting as it is taught in advanced classes. Suppose an attacker thrusts a long stick at your midsection, advancing with his left leg during the lunge and thrusting the stick horizontally, holding it with the palms of both hands down. It would be futile to try to stop the end of the stick. You rapidly step forward with your right foot so that the stick passes you on your left. (The agility to do this comes only with long practice.) As the stick passes you turn your body to face it so that you can grab it with both hands. Your left hand is forward of your attacker's outermost hand. Your right hand is between his hands.

In grabbing the stick your objective is not to stop its motion, which would require considerable force. Rather it is to deflect the lunge upward, around to your left in a circular motion and then up and over your attacker's head. Once he has committed himself to the forward lunge he can do little to prevent the deflection. He would need a large force to stop his momentum, and he cannot thrust horizontally to your midsection while pulling downward to prevent your deflection.

Once you have the stick over your attacker's head he is easily thrown. With his left foot forward he is highly unstable against a pull to his left rear because in that direction his center of mass must move only a short distance before it is no longer over his support area. When you have the stick over his head, you pull it downward over his back in that direction. He falls to the mat on his back and probably releases the stick.

Suppose you have a stick and a determined attacker rushes forward to grab its forward end. Allow him to grasp it but lead him with it (as if it were a carrot in front of a donkey) so that he continues his rush. Also lower your end to trick him into bending downward. Once he has committed himself to this awkward motion and is about to pass to your right you bring your end of the stick upward over his face and then downward over his back. If this motion is executed rapidly, he still has a strong grip on the stick and therefore is bent back-

ward by your pull downward over his back and by the continued forward motion of his torso. The torque due to his own weight rotates him to the floor around the pivot point of his feet. His

grip on the stick also provides a torque that rotates him. He actually throws himself to the mat because of his initial forward rush and a bit of trickery on your part.

Aikido has hundreds of techniques for employing such trickery against a determined opponent. In nearly all of them a small deflection force parries an opponent's thrust and then guides it so that he throws himself down. When I watch an aikido master defend himself, the motion seems so fluid and effortless that I am inclined to suspect the opponent of faking when he falls to the floor. The fall is not faked. It looks that way because the master has spent years developing an intuitive feeling for the basic physics of forces, rotation and torques.

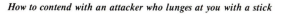

How to contend with an attacker who lunges at you with a stick

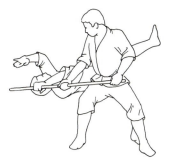

What to do when your opponent tries to seize your stick

5 Ballet

The essence of ballet maneuvers is physics

Ballet is a blend of beauty and the physics of motion. A man who is interested in both components is Kenneth Laws, professor of physics at Dickinson College and a student of ballet with the Central Pennsylvania Youth Ballet. Much of the following analysis of certain positions and movements of ballet is based on information that he provided me with.

A large part of the early training of a ballet student is aimed at teaching her (or, to be sure, him) to maintain her balance as she moves gracefully through the ballet forms. Balance is achieved when an area of support lies in an appropriate place below the dancer's body. If the support area is off to one side, gravity pulls or turns her toward the floor.

Gravity, of course, pulls constantly

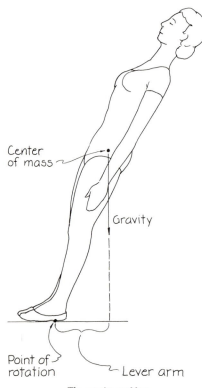

Center
of mass

Gravity

Point of
rotation — Lever arm

The gravity problem

on every part of the body, but the concept of a center of mass of the body helps to simplify the picture. The center of mass is a mathematical point whose position is determined by the distribution of mass within the body. The combined pull of gravity on all the parts of the body is said to operate through the center of mass.

When that center is not over a support area on the floor, the pull of gravity creates a torque on the body. A torque is the product of a force (here gravity) and a distance called the lever arm. Lines representing gravity and the lever arm appear in the illustration on this page, which shows a tilting human figure. The torque arising from the tilt tends to rotate the body about the feet and toward the floor. The greater the tilt, the longer the lever arm and the stronger the torque. If the dancer stood upright, the length of the lever arm would be zero and gravity would create no torque; the position would be stable.

The routine for the beginning student seeks to develop a sense of balance gradually. For example, she learns the position called the arabesque in a simple first version and is then advanced to the first arabesque penchée. In the simple version (arabesque à terre) she puts her right leg forward and her left leg back with its toes touching the floor; her weight is therefore on the right leg. The right arm reaches forward, the left arm slightly back.

Moving the left leg to the rear shifts the center of mass in that direction. Without a compensating movement the dancer would fall over backward. To shift the center of mass back over the support area of the right foot she must lean forward and extend her right arm. This motion serves two purposes: it is graceful and it enables her to maintain her balance.

In the first arabesque allongée the dancer continues the motion until her torso and right arm are almost horizontal and her rear leg is tilted upward. In the first arabesque penchée her torso and right arm are tilted downward and

her rear leg is tilted upward by 45 degrees or more. In both forms of the position the mass moved rearward by the rear leg must be matched by mass shifted forward by means of leaning the torso and extending the arm. Only then will the center of mass remain over the support area so that the dancer is stable.

Some of the fascination of ballet stems from illusions in which physical laws seem to be momentarily suspended. An example is provided by the grand jeté, a forward leap. According to Laws, a properly executed grand jeté suggests that the gravitational pull on the dancer somehow weakens when she is near the top of the leap.

Two things contribute to the illusion. First, some slowing does take place near the top of even an ordinary jump because of the rules governing motion. Although the dancer's horizontal speed remains constant throughout the jump, her vertical speed is zero at the peak. Just before and just after that point her vertical movement is slow. As a result for about half of the time required for the entire leap she is within a fourth of the leap's maximum height.

In a grand jeté another element is added to enhance the illusion that the dancer is hovering near the peak. Because of a certain shift of her arms and legs while she is in the air her path appears to be flattened at the top. The illusion depends on a shift of her center of mass as she moves her arms and legs. She launches herself with her arms downward. As she approaches the top of the leap she lifts and spreads her legs and arms. Her center of mass is then at a higher position in her body. Since the center of mass follows a fixed trajectory, the shift means the head and torso do not rise as far above the floor as they would have risen. In descending the dancer lowers her legs and arms, restoring the center of mass to its usual position.

In jeté en tournant, a turning leap, the dancer jumps into the air with no apparent spin around her vertical axis, yet near the top of the leap she begins to rotate. Impossible. One of the firm rules

First arabesque
à terre

First arabesque

First arabesque
penchée

Three forms of the arabesque

1   2   3   4

The grand jeté

Path of head

Raised
center of mass

Path of
center of mass

Initial center
of mass

How the dancer's center of mass shifts in the grand jeté

Small moment of inertia, large spin rate

The jeté en tournant

of physics is that the angular momentum of an object remains constant unless a torque acts on the object. If the dancer was not spinning when she left the floor, she cannot begin to spin in midair.

The explanation of the illusion is that the dancer does have a small spin at the beginning of the leap because of a torque she receives from the floor as she launches herself into the air. The spin is too slight for an observer to notice it. As the dancer rises, however, she pulls in her arms and brings her legs together. The effect is to decrease her moment of inertia. Since her angular momentum is fixed once she is in the air, the decrease in her moment of inertia increases her spin rate. One sees the same thing when a spinning figure skater pulls in her arms and so spins faster.

The dancer begins a jeté en tournant with her feet in what is called the fifth position. Her feet are parallel and point-

The grand jeté en tournant entrelacé

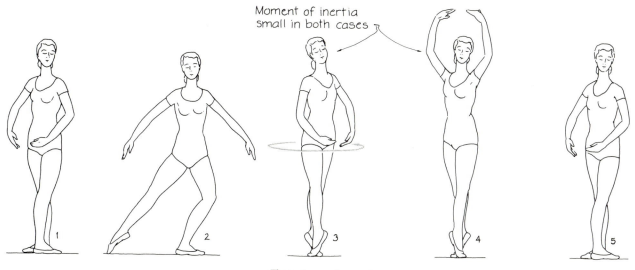

The soutenu en tournant

ing to the side in opposite directions, with the left foot in front. Each heel is near the toes of the other foot. The dancer slides the left foot to the left and quickly comes into an arabesque. She bends her left knee in what is called the demi-plié position.

Now she quickly brings her right leg downward and forward, simultaneously launching herself with her left leg. In flight she not only rises above the floor but also travels horizontally. At the beginning of the leap her arms and one leg are extended, providing the basis for her subsequent movements to increase her rate of spin.

The dancer lands on her right leg and in demi-plié. She should be facing the audience. With experience she learns to control the leap and her moment of inertia to achieve this orientation.

An equally beautiful but more difficult leap is the grand jeté en tournant entrelacé (also called the grand jeté en tournant or simply the tour jeté). The dancer starts with her right leg in demi-plié and her left leg lifted to the side at an angle of about 45 degrees with the floor. Next she steps backward onto her left leg, bringing her right leg upward and around to the front so that it points to her left just as she turns her body to the left. She jumps from the left leg. Once she is in the air she rotates about an axis that tilts from the vertical. She holds her arms above her head and close to the rotational axis, also bringing her left leg up near her right leg so that they both rotate about the axis. She lands on her right leg and comes into the first arabesque.

The rotation of the dancer in midair is rapid because her moment of inertia is small. As soon as she leaps she brings her arms and legs into line with the rotational axis. Once she has made her turn she effectively stops her rotation by spreading her arms and extending her right leg for her landing.

Classical dance has many techniques for whirling. One of them is embodied in the soutenu en tournant. Starting with her feet in the fifth position, left foot in front, the dancer goes into a demi-plié with her weight on her left leg and her right leg extended to the side. She brings her right foot back to the fifth position in front of the left foot and rises on her toes, twisting her feet to develop a torque that begins her twirl to the left.

A pirouette is a more ambitious twirl in which the propulsion also comes from torques on the feet. Consider first a quarter-turn pirouette. From the fifth position (right foot in front) the right foot is moved to the side while the arms are brought forward and then spread out to the sides. Next the right hand is moved to the front, the right foot to the rear. Now the dancer pushes against the floor with her right foot to propel herself around to the right. Simultaneously she rises on her left foot, which acts as a pivot.

The rotation brought about by the torque from the right foot is helped by the fact that the dancer draws in her left arm. The movement not only adds grace

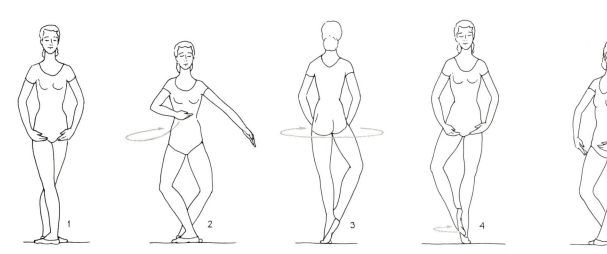

Portions of a pirouette en dehors

to the turn but also decreases the moment of inertia, enabling her to do the quarter turn rapidly. After dropping back into the fifth position she can immediately begin another quarter turn.

A complete pirouette differs mainly in the motion of the head and in the strength of the angular acceleration. A larger torque is required for a full turn. At the beginning of the rotation the dancer continues to face straight ahead. When her body has turned about 90 degrees, she suddenly whips her head in the direction of the turn, bringing it sharply to the front again when her body has turned about 270 degrees. Soon thereafter her body is also facing forward once more.

A grand pirouette calls for the dancer to turn with one leg and both arms held horizontally to the side. A mathematical analysis of this pirouette is difficult unless the shape of the human body is simplified. Laws has done so in a model that appears in the illustration at the right. The "body" consists of an upper section of mass M and length L and a two-part leg section. The "shank" and the "foot" have a combined mass of $m/3$ and a length of $L'/2$; the "thigh" has the same length but a mass of $2m/3$. For a male dancer the ratio $M:m$ is about 3.8 and L' is approximately equal to L.

The upper body is vertical throughout the maneuver. In the model of the pirouette one leg is held horizontally to the side. The supporting leg is rigid and makes an angle theta (θ) with the vertical axis through the supporting foot.

Laws was interested in the value of the angle required for a stable spin around the vertical axis. He first calculated it for a stationary dancer. As I have mentioned, the requirement for stability is that the dancer's center of mass be over the point of support. Since one of her legs is held horizontally to the side, she must tilt her other leg in order to shift mass and so reposition the center of mass. Stability is achieved when the angle between the supporting leg and the vertical is about 4.4 degrees.

Is it the same when the dancer is rotating? I would have thought so, but Laws's calculations indicate that the angle is instead about 3.5 degrees. Rotation imposes additional constraints on stability. The upper body must be closer to the vertical axis and the extended leg must stretch farther from the axis. Hence the dancer's center of mass is slightly off the vertical axis, and yet she is stable.

A more complicated rotation is part of the fouetté turn. (In Act III of *Swan Lake* the Black Swan does 32 consecutive turns.) The turn is essentially a full pirouette except that the source of the torque on the dancer is hidden. She continues to turn as though she were a toy top propelled by magic.

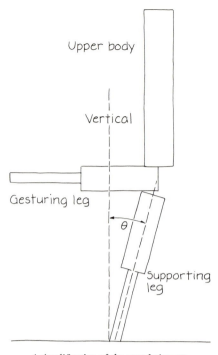

A simplification of the grand pirouette

Once the pirouette has been started the right leg is not returned to the floor until the end of the turn. During most of the pirouette the right foot is held near the knee of the supporting left leg. The dancer is on the toes of her left foot. Just as her body begins to face forward again she thrusts her right leg forward and opens her arms to the audience. The heel of the left foot is brought down to the floor and the left leg is bent. Then the right leg, still pointed outward, is brought around to the left side, continuing the rotation, while the rest of the body remains facing the audience. When the right leg has been thus rotated through about 90 degrees, the right foot is brought back to the left knee and she comes back en pointe (on her left toes). She then repeats the entire procedure.

The key question is how the dancer generates enough torque to continue the rotation. With each turn the friction between the supporting foot and the floor robs her of angular momentum and spin. The secret is in the movement of the right leg. When it is brought forward and rotated to the right, it takes up whatever angular momentum the dancer still has. Her rotation stops, except for the right leg, giving her a moment to put her left foot flat on the floor. In this moment she can push against the floor, providing the torque for another rotation. To assist it she draws her right foot toward her left knee to reduce her moment of inertia as she spins.

The fouetté is difficult for dancers. Laws points out that a novice is likely to botch the turn by thrusting the right leg directly to the side instead of positioning it so that it absorbs the angular momentum properly. The novice might also extend the leg only partly forward toward the audience. It must be fully extended to absorb all the angular momentum.

The moment of inertia of the extended leg is about 1.7 times more than the body's. Hence when the leg absorbs the body's angular momentum, it does not turn as rapidly as the body. If the normal pirouette is at the rate of two revolutions per second, the extended leg turns at only 1.2. If the leg absorbs all the dancer's angular momentum, she has about .3 second to push off for another turn.

In the grand pas de chat the application of rotational dynamics is needed to enable the dancer to leap while she maintains the orientation of her body. From the fifth position of the feet she does a demi-plié on the right foot, bringing the left leg behind the right one at an angle of about 45 degrees from the vertical. The jump is made from the supporting leg. Once the dancer is in the air she brings the right leg into line with the left leg.

This alignment of the legs requires a rotation of the left leg about the dancer's center of mass. Yet her angular momentum in midair is essentially zero. How does she manage a rotation while keeping her angular momentum constant and maintaining the orientation of her torso? She rotates her arms in the direction opposite to the movement of her left leg. The arms are moved just enough to make the combined angular momentum of the arms and the right leg zero. By means of these countering rotations the dancer can rotate part of her body in the air.

Laws has analyzed the subtleties of balance achieved in the promenade en attitude derrière. Here a female dancer is in the position derrière en pointe: raised on the toes of one foot with the other leg poised in the air. One of her hands is held by a male partner, whom she faces. The other hand is raised in a graceful arc. The sustained pose is difficult because of the balance required. If the female dancer shifts into this pose from another movement, she is likely to be off balance.

The dancer could correct her balance by stretching her body in order to shift the center of mass back over the supporting foot, but that would disturb the grace of the movement. She could also push or pull her partner's hand, but that would be likely to generate a torque that would make her turn on her supporting foot, rotating her away from her partner when she should face him.

A better method is to achieve the proper application of forces between the touching hands. According to Laws,

A fouetté en tournant with two discrete turns

Rotation of leg and arm
in opposite directions

The pas de chat

both dancers' hands should be horizontal and the elbows should be raised. Then the female dancer can apply forces to her partner's hand in such a way that a turning torque does not develop. The forces on her hand are indicated in the illustration on page 32. Two oppositely directed forces come from the male dancer's hand, each one displaced from the center of the hand by a distance d. The center of the hand is at a distance D from the vertical axis running through the female dancer's supporting foot.

The forces on the hands generate torques on the female dancer, but they

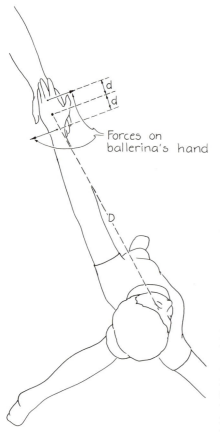

Forces on
ballerina's hand

The balancing forces in derrière en pointe

tend to turn her in opposite directions. By properly controlling the forces she can make the two torques cancel. She makes the nearer force slightly larger than the farther one. Since torque is the product of a lever arm and a force and the lever arm to the nearer force is smaller, the torques cancel. The forces on her hand therefore do not cause her to rotate about her pointed toes. By twisting her hand in just the right way, however, she can move toward or away from her partner's hand, thereby adjusting the position of her center of mass in such a way that the audience does not realize what she is doing.

Several of the male dancer's leaps call for his beating his calves together while he is in flight. One is entrechat quatre, which begins with the feet in the fifth position. The dancer descends into a demi-plié and then propels himself directly upward. In the air he opens his legs to the side, brings them briskly together, separates them and then brings them together again as he lands in the fifth position demi-plié. An experienced dancer with a powerful leap might be able to complete two beats or more.

Laws has investigated the difficulty this movement presents for a large dancer. In general such a dancer cannot beat his legs together at the rate and amplitude that can be achieved by a smaller dancer. Even though the larger dancer may be stronger, his legs are more massive and therefore more difficult to rotate. He also has more mass to lift and more difficulty leaping to the same fraction of his height.

I have only touched on the movements of classical dance. You could do a good deal more. Try analyzing a ballet movement. It would be helpful to make a series of photographs with stroboscopic flash. A motion picture that can be run slowly or stopped periodically would be even better. None of these techniques will be easy because the shape and motions of the human body are complex. You might simplify the shape of the body as Laws has done in some of his analyses.

A study of ballet is also complicated by the requirement of grace and style. If you want to determine the physical principles underlying certain movements, you must distinguish the components of the movements that are done only for style. Both Laws and I shall be interested in hearing about what you find out.

NOTES

Kenneth Laws has recently published a beautiful book titled *The Physics of Dance* (see the Bibliography) in which he analyzes ballet in far richer detail and with more understanding of the grace of ballet than I have. His book is also better at naming the various moves of a ballet performer.

Several people pointed out a restriction on my claim that ballet performers cannot initiate rotation while in the air. A more rigorous statement is that they cannot initiate any net rotation of their body once they have left the stage. However, they can rotate part of their body one way if another part rotates the opposite way. Cats undergo such rotations when they right themselves after having been dropped inverted. Springboard divers and gymnasts also undergo such gyrations. Laws examines an example (demi-fouetté) of such a maneuver in his book.

6

Rattlebacks and Tippe Tops

The mysterious "rattleback": a stone that spins in one direction and then reverses

If you spin a certain type of stone in the "wrong" direction, it will quickly stop, rattle up and down for a few seconds and then spin in the opposite direction. Going in the "right" direction, however, it will usually spin stably. The stone is apparently biased toward one direction of spin. It will even develop a spin in that direction if you just tap one end downward. The rocking of the stone caused by the tap is quickly converted into a spin.

These curious stones were originally called celts because their behavior was discovered by archaeologists studying the prehistoric axes and adzes so named. I imagine the spin reversal was first noticed by an archaeologist idly spinning one of the celts on a table. My introduction to the stones came some years ago when I read Harold Crabtree's delightful book *An Elementary Treatment of the Spinning Tops and Gyroscopes*. Among the many spinning tops, hoops and other toys he described were the mysterious celts, but he did not indicate why they exhibited a bias for one direction of spin. Recently two people have rekindled my interest in the stones. Nicholas A. Wheeler of Reed College has been considering doing an analytical investigation of them. (If he does one, it will be to my knowledge the first study of the stones since the 19th century.) A. D. Moore of the University of Michigan has sent me several of the hundreds of celts he has made out of dental stone, spoons and other materials. Moore has dubbed his stones rattlebacks, a term I shall employ here.

Most rattlebacks have a smooth, ellipsoidal bottom. The top can be flat, hollowed out or ellipsoidal. The spin bias apparently results from the shape of the bottom and the distribution of the mass with respect to the axis of spin. The basic rattleback in Moore's collection has a flat rectangular top. The key feature of the design is that the long axis of the ellipsoid is aligned at an angle of from five to 10 degrees to the long axis of the rectangular top. This skewing (and per-haps some subtle shaping of the ellipsoid) introduces the bias to the spin of the stone but in a way that may be difficult to see in detail.

Most of the rattlebacks Moore has sent me spin clockwise as seen from above. When I spin such a rattleback clockwise, it rotates rather smoothly until friction eventually stops it. When I spin it counterclockwise, it first turns through a few revolutions but then stops as its ends begin to rock up and down. Soon it starts to spin clockwise until friction halts it. If I tap the end of the stone when it is stationary, the ends chatter up and down for a few seconds and then the stone begins to turn clockwise. Again it is friction with the table that eventually stops the stone.

Some of the rattlebacks display a second reversal, from clockwise to counterclockwise, near the end of the spinning just as the friction is about to remove the last of the stone's energy. The second reversal is usually not as strong as the first and does not display the same type of vertical oscillation. Instead the stone rocks from side to side rather than longitudinally. A few of the rattlebacks display strong reversals with either direction of spin.

Moore forms most of his rattlebacks with dental stone, which he polishes and varnishes. Once he has made a good one (a feat that requires both experimentation and luck) he makes a mold of it for duplication. Some of his rattlebacks are made from the bowl of a spoon that he has detached from its handle. The bowl is glued to a flat rectangular piece of metal or Plexiglas with the long axis of the bowl's ellipsoid at an angle to the long axis of the rectangle. Pennies are glued on top of some of the rectangular pieces to add weight and to change the rattleback's moment of inertia.

One type of rattleback looks like half an egg. Across the flat top Moore tapes a brass rod at an angle to the long axis. Another type is similar to the basic design except that the inside is hollowed out to resemble a boat. Still another type has central terraces built on top of a rattleback of the basic design.

Moore's favorite rattleback must certainly be the half egg. He once engaged in a contest with C. L. Stong, who conducted this department for almost two decades, to determine which of them could get the most back turns out of a rattleback. Moore won with the half egg, which made more than 15 back turns before it stopped.

When I examined the rattlebacks, I found that the basic kind exhibited the strongest reversal (counterclockwise to clockwise), which is followed by a slight second reversal as the stone nears the end of its energy. The rattleback with the pennies and the one with the terraces performed about as well as the basic type. The half egg did not spin well, tending to wobble so much that the ends of the brass rod scraped the table and thus drained energy from the egg. Moore is not sure why this former winner now does so poorly, but he suspects that a hairline crack on the bottom surface may be the cause.

The most fascinating rattleback is the boatlike one because it displays strong reversals in each direction of spin. When I spin it counterclockwise, it rotates for part of a revolution, stops, wobbles briskly up and down and then reverses for several revolutions. Then it stops spinning again, rocks from side to side and begins to spin counterclockwise once more. The performance continues until the stone runs out of energy.

What causes the reversals of spin? Although a complete mathematical explanation is difficult, careful observation reveals a few clues. When a rattleback of the basic design begins to reverse from a counterclockwise rotation, its longitudinal oscillations appear to be approximately around the short axis of the ellipsoid because the points marked C and D in the illustration on page 36 vibrate up and down more than the points marked A and F. (I think the oscillation is actually around another important axis, one of the stone's prin-

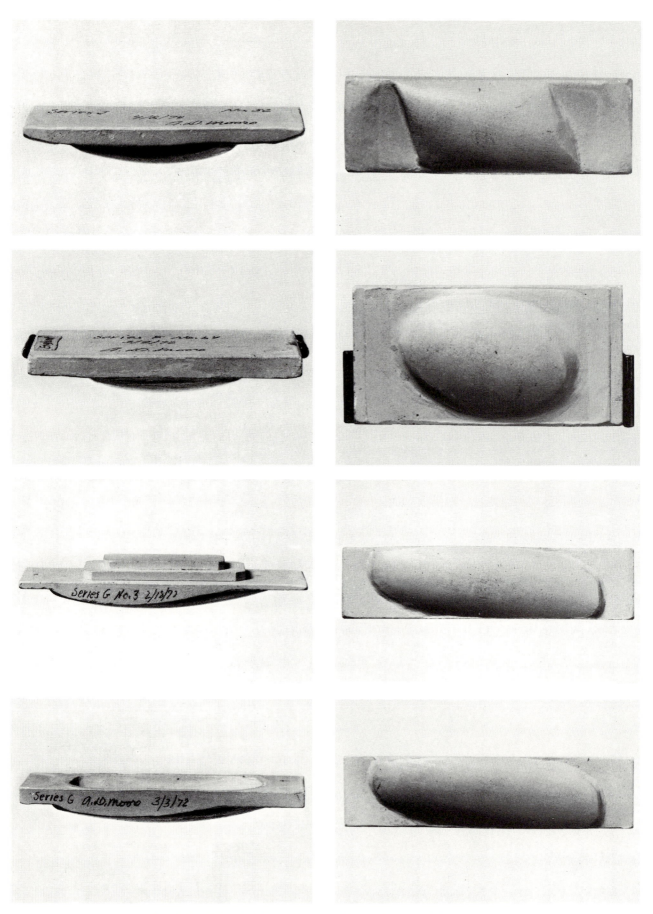

*Paired views of the side (**left**) and bottom (**right**) of rattlebacks made by A. D. Moore*

cipal axes that is almost parallel to the short axis of the ellipsoid.) Somehow the counterclockwise rotation excites an instability of this kind.

When the rattleback executes its second and milder spin reversal, it appears to oscillate approximately around the long axis of the ellipsoid, sending points *G* and *H* up and down with more amplitude than other points. (Again, theoretically the actual axis is a principal axis closely aligned with the long axis.) Apparently the clockwise spin excites this type of instability.

I can easily set up either type of oscillation in an initially stationary stone by pressing down on an appropriate part of the stone and then letting go. When I do this at point *A,* not much happens. The rattleback oscillates for a short time and may turn counterclockwise slightly. At *B* a similar action results in more pronounced oscillations and a clockwise rotation, but the performance is still far from dramatic. Livelier effects are obtained with a press and release at point *C;* the longitudinal oscillations are vigorous and the stone is quickly propelled into a clockwise rotation. The oscillations are identical with the ones I see when the spin reverses from counterclockwise to clockwise.

When I press point *G* and release it, the stone rocks from side to side and soon begins to spin counterclockwise. Thus when I excite the vertical oscillations at *G,* I am duplicating the second type of spin reversal (of the initial clockwise rotation). Similar results are obtained with a press and release at other points between *G* and *A.*

The transverse oscillations are much slower than the longitudinal ones, but both types are so fast that I cannot follow the motion easily. Moreover, once the oscillations start the stone begins to turn and my perspective is altered. To follow the oscillations better I tried tapping a rattleback I had placed on a strip of tape that had its sticky side up. The tape kept the stone from rotating, but of course it may also have interfered with the stone's natural motion. Although the stone seemed to oscillate primarily around a single axis, it also gave the appearance of rolling out of the primary plane of oscillation. I do not know if such a feature is essential to the reversals of spin.

The first scientific investigation of spin-reversing stones was made by G. T. Walker (no relation) in 1896. Some of his results can still be found in reprints of old books dealing with rotational mechanics. Usually the description of the stones is put in the chapter on asymmetrical tops, that is, tops that are not symmetrical around the vertical axis. Theoretical discussions of symmetrical tops are challenging enough, but the theo-

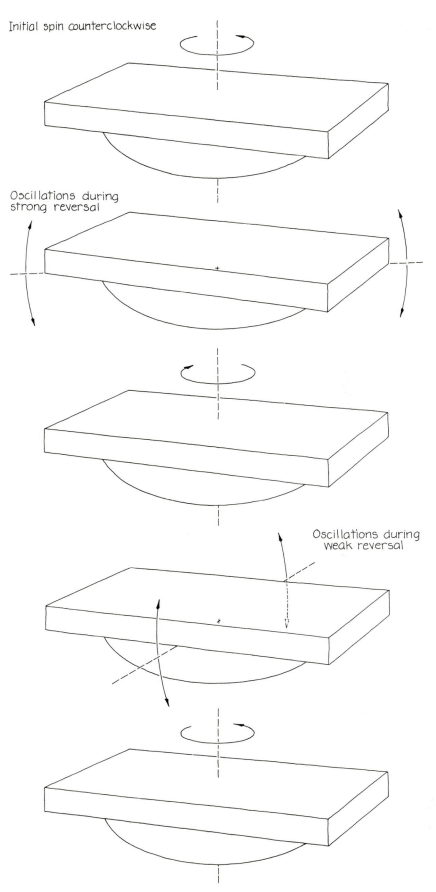

Initial spin counterclockwise

Oscillations during strong reversal

Oscillations during weak reversal

A rattleback that exhibits a strong reversal and a weak one

ry of asymmetrical tops might be called mind-spinning.

A successful rattleback has several key features. One has to do with the misalignment of the stone's principal axes and the axes of the ellipsoid on the bottom. A principal axis of an object is an axis about which the object could be rotated freely with no rotation about either of the other two principal axes. The three mutually perpendicular axes are usually found along the axes of symmetry of many common objects. If the ellipsoid were aligned with the rectangular top in Moore's basic rattleback, the principal axes would be easy to find because they would coincide with the axes of symmetry. One axis would be vertical and the other two would be parallel to the short and long axes of the ellipsoid. They would all cross at the center of mass of the stone.

A good rattleback's ellipsoidal bottom is not aligned with the rectangular top. The vertical axis is still a principal axis, but the other two axes are now shifted around the vertical one in the direction in which the ellipsoidal bottom has been shifted out of alignment with the rectangular top. One of these principal axes, call it axis 2, is somewhere between the long body axis of the rectangular top and the long axis of the ellipsoid. The other one, axis 3, is somewhere between the corresponding two short axes and is perpendicular to axis 2. Their exact locations depend on the relative mass distributions of the rectangular top and the ellipsoid. The misalignment of the ellipsoid from the principal axes is an essential feature in the reversal of spin.

A second important feature of a well-trimmed rattleback is that the radius of curvature is different along the two axes of the ellipsoid. (A large radius of curvature corresponds to a surface with a small amount of curvature.) If the bottom were perfectly spherical, the rattleback would not reverse its spin.

It is also important that the mass distribution of the stone be different for the two principal axes. The function giving the distribution of mass with respect to a particular axis is called the moment of inertia. Consider principal axis 3. When one of Moore's rattlebacks oscillates up and down around that axis, as it does during the strong type of spin reversal, the moment of inertia for the oscillation is relatively large because some of the mass is at a relatively large distance from the axis. Suppose the stone oscillates around principal axis 2, as it does during the second, weaker type of spin reversal. The moment of inertia for that oscillation is relatively small because most of the mass is relatively close to the axis. This difference in the moment of inertia of the two horizontal principal

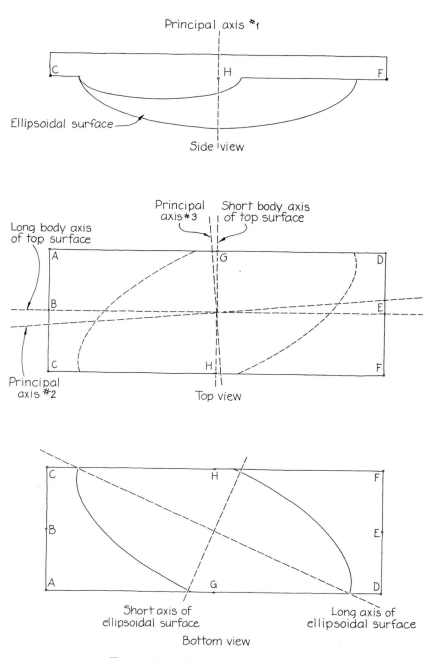

The axes of a rattleback biased to rotate clockwise

axes is essential to the success of a rattleback.

In sum, Moore's basic rattleback has the following key features. The long axis of the ellipsoid (the one with the large radius of curvature) is shifted counterclockwise (as seen from above) from principal axis 2, which has a small moment of inertia. This arrangement biases the rattleback for a strong spin reversal from counterclockwise to clockwise and a weaker spin reversal in the other direction. What would happen if the long axis of the ellipsoid were shifted in the other direction? Would the strong and weak spin reversals be in the opposite directions? Although G. T. Walker's theoreti-

cal work predicts the behavior, I shall leave the questions for you to answer with suitably carved rattlebacks.

How do the key features of Moore's rattleback cause the stone to reverse spin? Primarily they make it unstable against small perturbations of the spin around the vertical axis. If the bottom were spherical or the ellipsoid were aligned with the principal axes, any small perturbation would have a small effect on the spin and would not cause noticeable oscillations. With a properly shaped stone, however, small perturbations from the initial spinning of the stone or from the tabletop generate oscillations that rapidly grow in amplitude. Which of two types of instability is excited depends on the direction of spin and the design of the stone.

Suppose a rattleback of the basic shape is spun counterclockwise. Because of its design any small perturbation causing an oscillation around principal axis 3 is rapidly enhanced. The frictional forces acting on the stone during the oscillation stop the spin and then initiate an opposite one. Once the reverse spin has begun, the friction forces stop the oscillation. Hence a spin in the counterclockwise direction allows one type of instability—the oscillation around principal axis 3—to grow large quickly, and then the frictional forces accompanying the oscillations reverse the spin.

If the stone is initially spinning clockwise, another instability is encouraged, one that enhances oscillation around the other horizontal principal axis. Again the design of the rattleback allows this instability to grow in amplitude exponentially. Frictional forces during this kind of oscillation cause the clockwise spin to stop and then reverse.

I wanted to investigate how the behavior of a basic rattleback would change if I altered one of the key design features. The only feature I could change easily was the relative moment of inertia for the two horizontal principal axes. To alter it I taped a pencil across the top of a rattleback, first parallel to the length of the rectangle and then parallel to the width. Each time I was careful to balance the pencil so that the rattleback would spin on the same area on the ellipsoid.

With the pencil lengthwise the rattleback behaved generally as it had before because the pencil merely increased the difference in the moments of inertia around the principal axes. With the pencil parallel to the width the stone did not reverse spin or show any of the characteristic instabilities. By placing the pencil across the width of the rectangle I was increasing the moment of inertia around principal axis 2 without much changing the moment of inertia around principal axis 3. Apparently I increased the one enough to make it comparable to the other. When the moments of inertia are much the same, the oscillations due to small perturbations do not grow exponentially with time and therefore remain small. In the absence of large oscillations around the principal axes the frictional forces from the tabletop cannot stop the rattleback and then reverse its spin.

A moment of inertia can be varied in a more controlled way by mounting long stove bolts on a rattleback. Extend a bolt in each direction along a principal axis and screw several nuts on the ends of the bolts. The moment of inertia is changed by moving the nuts closer to the center of the rattleback or farther away from it. (The stone must still be balanced.) With this arrangement I was again able to eliminate the spin reversals when I equalized the moments of inertia for the two horizontal principal axes. What would happen if one adjusted the nuts to make the moment of inertia around principal axis 2 larger than the one around principal axis 3? Would spin reversal appear again but with the directions of the strong and weak reversals interchanged?

Moore's half-egg rattleback has a brass rod by which the moments of inertia can be altered. (Repositioning the rod also shifts the principal axes somewhat.) When I tape the rod across the short width of the flat top, the egg spins stably in each direction. With the rod along the longer axis of the flat top instabilities and spin reversals appear. Normally the rod is positioned at a small angle to the longer axis of the flat top, as is shown in the illustration below. The half egg reverses weakly in each direction of spin. With the rod mounted in this way the half egg is also sensitive to a tap on its perimeter. The direction of the resulting spin depends on the location of the tap. The illustration indicates the possibilities.

G. T. Walker's analysis predicts that the frequency of oscillation is higher around principal axis 3 than it is around axis 2, a point easily verified with a basic rattleback. His equations also predict that neither instability will arise if the spin frequency is higher than the oscillation frequency. With Moore's rattlebacks my attempt to eliminate the instability for the strong reversal (the one for a counterclockwise spin) is impractical because the vigor of my initial twist inevitably introduces instabilities. I can easily spin the rattleback clockwise, however, with a spin frequency greater than the oscillation frequency around principal axis 2. As far as I can tell the oscillation does not appear until friction has slowed the stone, presumably to a spin frequency lower than the oscillation frequency.

I played with Moore's rattlebacks on a smooth Formica top and a few other surfaces. I have not investigated the effects of the friction coefficient between the rattleback and the surface on which it is spun. If the friction is too high, the spinning should be eliminated so quickly that no spin reversal develops. If the friction is too low, the stone will reverse spin with difficulty because the reversal requires friction.

You can easily carve rattlebacks from dental stone or wood. Adjust the shape of the bottom until the rattleback works properly. You may be able to think of variations of the basic design. For example, you could investigate the effect of weight distribution by taping small weights or metal rods on the top of the rattleback. By making a number of rattlebacks you could determine how the orientation of the ellipsoid on the bottom affects the spin reversal. If you are out to win a contest for the number of spin reversals of a rattleback, look for an optimum angle between the long body axis and the long axis of the ellipsoid. You might also try to build a rattleback big enough to ride on. If you could make one with a vigorous spin reversal, riding it would be like riding a bucking bronco.

I turn now to the Tippe Top, which also reverses unexpectedly. Such a top usually has a spherical bottom and a central stem. It is spun by sharply rolling the stem through the fingers and dropping the top on a flat surface. The sur-

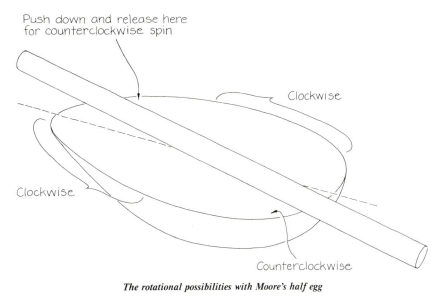

The rotational possibilities with Moore's half egg

prise is that the top spins on its spherical end for only a few seconds and then, turning upside down, spins on its stem. The motion appears to violate the law of the conservation of energy because the top seems to raise its center of mass (which is in the spherical section) without outside help.

The top has long fascinated observers, including several distinguished physicists and mathematicians. In a recent paper Richard J. Cohen of the Massachusetts Institute of Technology describes how William Thomson (the eminent physicist better known as Lord Kelvin) spent his time spinning smooth stones on the beach instead of preparing for his mathematical examination at the University of Cambridge. Later Niels Bohr, who developed the first modern model of the hydrogen atom, became similarly entranced with the mechanics of the Tippe Top.

The surprising inversion of the top arises from friction between the top and the surface on which it spins. Suppose it spins with its stem tilted away from the vertical axis, as is shown in the illustration below. Because it is spinning it develops a sliding friction force (perpendicular to the plane of the page in the illustration). This force creates a torque about the center of mass, causing the top to invert.

You could make a Tippe Top by cutting off part of a solid rubber ball and inserting a bolt in the flat surface. A typical school ring with a smooth stone displays the same type of inversion. Spin the ring on the stone. A hard-boiled egg that is spun flat will rise to spin on one end. Indeed, spinning an egg is one way to ascertain whether it is hard-boiled; a fresh egg will not stand up because of the sloshing of its internal liquid.

NOTES

Rattlebacks show up in odd places. Some people wrote me about how they found natural stones that displayed the spin reversals of rattlebacks. Even some types of pocket knives behave this way. Allan Adler of Brandeis University pointed out that some stones of the Olmec Indians will undergo reversal of spin. Whether the stones were designed for such play or merely turned out that way is unknown.

Two modern studies of rattlebacks have now appeared and are listed in the bibliography. T. K. Caughey of the California Institute of Technology developed a mathematical model that displays most of the features of a rattleback's spin reversal. Thomas R. Kane of Stanford University and David A. Levinson of the Lockheed Palo Alto Research Laboratory constructed a more rigorous mathematical model of the rattleback. They found that the instability leading to spin reversal is due to the offset of the body axes and the principal axes, as I had guessed. What I did not appreciate before their work is that the rattleback wobbles as it spins. This additional motion allows it to roll over the table top instead of slipping.

If there is no slipping, then the rattleback should not lose energy to friction. Why then does it stop? Kane and Levinson discovered that the principal loss of energy is to air drag. When they mathematically followed the motion of a rattleback with no energy losses to air drag, the rattleback behaved in an ideal manner. Suppose their model rattleback is like one of Moore's in that it prefers the clockwise sense of spin. When the mathematical model is spun counterclockwise, it soon reverses direction and spins clockwise for a longer time. Then it reverses direction again. Thereafter, additional reversals occur approximately periodically.

When Kane and Levinson added an appreciable air drag to their calculations, the rattleback was limited to the first reversal. When the rattleback was initially spun in the clockwise direction, it never reversed. Thus air drag does seem to be important to the behavior of a real rattleback. You might like to test this conclusion by arranging to spin rattlebacks in a vacuum.

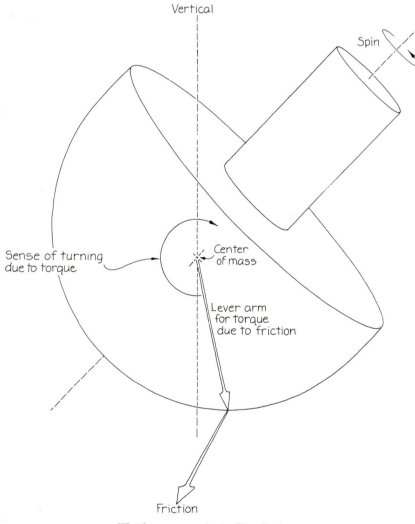

What happens as a spinning Tippe Top inverts

Tops

7

The physics of spinning tops, including some far-out ones, and more on the tippe top

Spinning tops are ancient toys, but the principles that govern their behavior have come to be understood only in the past century and a half. What keeps an irregularly shaped object spinning on a single point? Why do tops of different shapes behave in such different ways? Here I shall explain some of the mechanics of tops, avoiding the mathematical thickets that obscure the subject in some physics textbooks. I shall also introduce several of the unusual tops made by Donald W. Dubois of the University of New Mexico, who has set spinning such unlikely objects as a golf tee and the stopper from a bottle of India ink.

The behavior of any top is due mainly to the effect of gravity. Every atom of the top's mass is pulled downward by gravity, but the net pull is more readily envisioned as being through the center of mass, which lies somewhere inside the top, usually at its geometric center. The weight of the top can be represented by a vector pointing downward from the center of mass. It seems logical that because of this pull the top would be less likely to remain standing than to topple over (as you would if you were leaning away from the vertical).

The difference is that the top is spinning. As a result the downward pull of gravity gives rise to the surprising rotation of the top about the vertical. This reaction is not easy to visualize because you are more familiar with nonspinning objects. Normally a force on an object causes an acceleration in the direction of the force. When a spin is involved, the force may result in a motion perpendicular to the direction of the force. Such

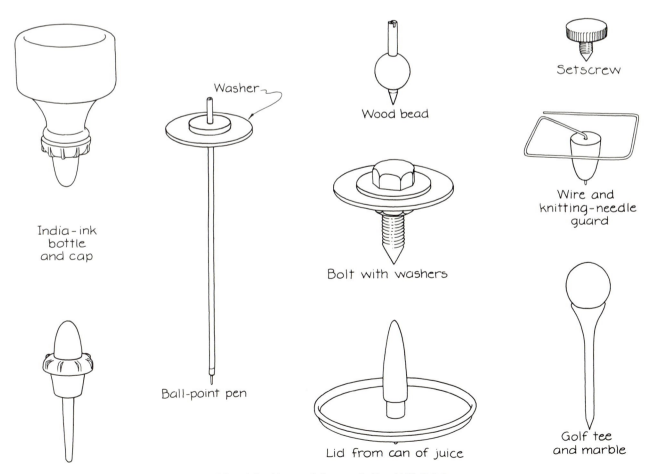

A few of the objects made into tops by Donald W. Dubois

India-ink bottle and cap

Washer

Ball-point pen

Wood bead

Bolt with washers

Lid from can of juice

Setscrew

Wire and knitting-needle guard

Golf tee and marble

an unfamiliar motion is part of the fascination of tops.

The spinning top has angular momentum. Where linear momentum is the product of the mass and the velocity of an object, angular momentum is the product of the mass distribution (the moment of inertia) and the angular velocity. The top has angular momentum because it is spinning around its long body axis and because it has a mass distributed around that axis. The angular momentum is a vector that lies along the body axis, the axis around which the top is symmetrical.

The only way to change the angular momentum of an object is by means of a torque. A torque is scientifically defined (not too differently from the common usage) as the product of (1) the force on an object and (2) a lever arm that extends from a pivot point to a line drawn through the force perpendicular to the lever arm. The pivot point for a top is obviously at the pointed end touching the floor or a tabletop. The lever arm extends horizontally from that point to a vertical line drawn through the center of mass, which is where gravity is considered to be pulling on the top.

Gravity provides not only a downward force but also a torque to change the angular momentum of a top. The torque does so in a simple way: it redirects the angular momentum, rotating the vector about the vertical axis. Since the vector must continue to be along the body axis, that axis also rotates about the vertical in the motion called precession. It maps out a cone centered on the vertical. (For the present I shall treat the point of the top as being fixed where it touches the floor.)

If the top is spinning counterclockwise as it is viewed from overhead, the precession around the vertical is also counterclockwise. If the top is spinning clockwise, the precession is clockwise. If the top were not spinning, the torque effect of gravity would cause it to fall to the floor, which it actually does during the last stage of its spin because of the effect of friction on the point on which the top is spinning.

Nearly all tops fall somewhat as they start to spin. The reason is an energy requirement. If the top is to process as a result of the torque due to gravity, it must have kinetic energy. Occasionally it gains energy from the launching. More often the energy must come from the initial fall, during which the decrease in the potential energy of the top is transformed into kinetic energy.

This simple story of the top fails the first time you examine the motion of a real top. The top does not just spin and precess; its body axis nods in what is called nutation. As it nutates the angle between the body axis and the vertical varies between two values determined by the mass distribution and kinetic en-

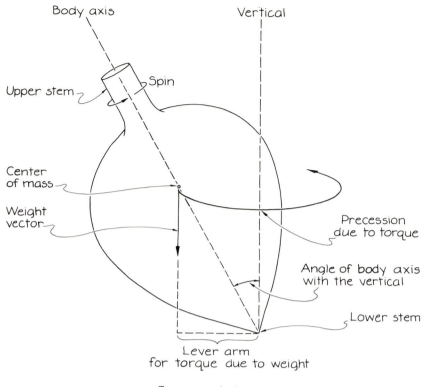

Torque on a spinning top

ergy of the top and its initial angle with the vertical. It is possible to illustrate the types of nutation by tracing the position of the upper end of the body axis on a sphere centered on the point where the top touches the floor. The angle between the body axis and the vertical is limited by two circles drawn around the vertical. The circles represent the range through which the top can lean away from the vertical during nutation.

In one type of nutation the body axis weaves between the two limiting circles harmonically, touching each circle tangentially. The precession of the axis around the vertical is always in one direction, either clockwise or counterclockwise as seen from above. In the illustration on page 41 the rotation is counterclockwise.

In the second type of nutation the body axis loops between the two limiting circles but still touches each one tangentially. The direction of travel of the body axis periodically changes between clockwise and counterclockwise. In spite of this reversal the precession has an average value that is in one direction or the other. In the illustration the average precession is counterclockwise.

The third type of nutation traces cusps on the imaginary sphere. The path of the body axis meets one of the limiting circles tangentially as it does in the other types. The path meets the other limiting circle perpendicularly. The precession is consistently in one direction, but the rate of precession varies from maximum at the lower limiting circle to zero at the upper limiting circle.

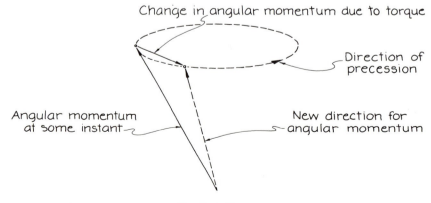

The effect of torque

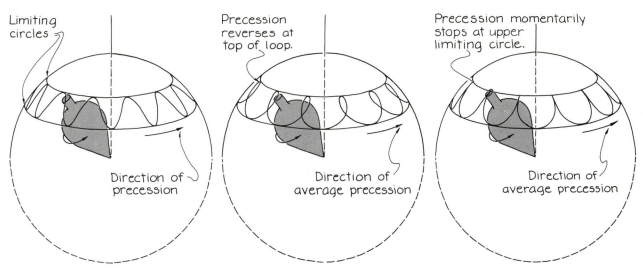

Limiting circles

Direction of precession

Precession reverses at top of loop.

Direction of average precession

Precession momentarily stops at upper limiting circle.

Direction of average precession

The nutation patterns of tops

The type of nutation that occurs (if any) depends on the initial conditions of the top's spinning. Since there are many perturbations in starting a top, you do not always have full control over those conditions. Suppose the launching imparts to the top a precession velocity that is in the same direction as the precession velocity gravity provides. Then the top precesses with harmonic nutation. Regardless of where the top is during its cycle of nutation, either its initial precession velocity or the velocity provided by gravity guarantees its continued precession in one direction around the vertical.

If the initial precession velocity is opposite to the precession resulting from gravity, the nutation is looped. During the lower part of the loop the gravitational precession drives the top around the vertical in the preferred direction. During the higher part of the loop, where the body axis reaches its upper limiting circle, the precession due to gravity is exhausted and only the initial precession is left. Until the top can drop and be driven around the vertical again by gravity it precesses in the opposite direction.

The third type of nutation is often seen when a spinning top is initially held at an angle to the vertical and then released. It has no initial precession and so it just falls to the lower limiting circle. The fall is necessary because the resulting decrease in the potential energy of the top provides energy for the precession brought about by the torque exerted by gravity. When continued nutation brings the body axis back to its initial angle with the vertical, the top regains its initial potential energy. Precession then stops momentarily, since no energy is left for it. Afterward the top again falls to the lower limiting circle and precession continues until the top again is raised by nutation.

The limiting circles originate from three severe restrictions on spin, precession and nutation. The total energy (kinetic and potential) of the top must remain constant. (I shall discuss the effects of friction below.) The angular momentum along the body axis must also remain constant in amount, although its direction can change, because there is no torque along the axis to change it. Finally, the angular momentum along the vertical must remain constant for the same reason. (Once precession begins, the calculation of these two angular momentums becomes harder because the angular velocity then includes both the spin of the top and the precession velocity.) Because of these three strict rules the body axis is limited to a certain range of angles with the vertical. If the body axis dropped below the lower limiting circle or rose above the upper one, it would do so only by disobeying the rules. (Tops do rise above the upper circle but only because of friction.)

Most mathematical models of tops are devised with the simplifying assumption that the kinetic energy of the spin is much greater than the change in potential energy as the top weaves in nutation. Such a top is said to be "fast." With this assumption several features of the motion can be related to the rate at which the top spins. A lower rate increases the nutation weave but decreases the rate of nutation. The average rate of precession also depends on the spin, being faster with a slower spin. You can easily see these relations when you spin a top on the floor. As friction gradually robs the top of spin the precession rate increases and nutation becomes slower and more pronounced. Finally, just before the end, the top swings sluggishly up and down as it precesses around the vertical faster than ever.

The precession rate and the extent and frequency of nutation also depend on

the mass and shape of the top. In general a greater mass increases each of these components of the motion. (Of course, a more massive top is more difficult to spin at a given rate than a less massive one.)

If the top is spinning quite rapidly, the small amount of nutation that should be present might be eliminated by the friction on the point of the top. Then the top would appear to precess uniformly around the vertical without nutation. This uniformity, however, is due only to the intervention of the friction.

Truly uniform precession is possible if the top has coincident limiting circles. If it does, it must be released in such a way that its angle with the vertical corresponds to the angle of the limiting circles. Such a top does not nutate, since it can neither rise above the limiting circles nor fall below them. Since it does not fall, however, it has no way to convert some of its potential energy into energy for precession. Your starting technique or the initial impact of the top on the floor would have to provide the impetus to start precession.

There are actually two possible types of precession. Only the slower one is related to the gravitational pull on the top. The faster one is rarely seen in a top but can turn up under the proper initial conditions. The classical explanation of the faster precession involves a top (usually considered as an ellipsoid in textbook discussions) that is free to roll around on a horizontal plane while it spins. No forces or torques act on this imaginary top, not even gravity. In this situation the top cannot slide into the plane or break contact with it.

Once the top is spinning with its body axis at some angle to the vertical it begins to precess around the vertical in a way that maintains the contact between it and the horizontal plane. The plane is usually called the invariant plane; the path of the top's contact point on the invariant plane is called the herpolhode, and the path of the top's contact point on the top itself is called the polhode. As one writer put it, the motion leads to a jabberwockian statement: The polhode rolls without slipping on the herpolhode lying in the invariant plane. The precession in this idealized case is the fast precession a real top might have.

My colleague James A. Lock has pointed out that fast precession can be seen in a spinning football and a soft-drink bottle thrown into the air. A quarterback throws a football with a spin to keep the ball stable during flight, taking advantage of its streamlined shape. In a long pass the football often has a noticeable wobble, which may be partly due to air drag. Even without drag the ball will wobble if it is thrown to develop fast precession.

The precession is easier to see if a soft-drink bottle is tossed upward with spin. (The bottle should be empty; sloshing fluid would interfere with the spin.) If you toss the bottle with its body axis vertical, you may not be able to see any precession. With a less careful toss the spinning bottle will precess around the vertical in the fast-precession mode. The motion is not caused by gravity because the bottle is in flight. During the precession the bottom of the bottle traces out the herpolhode on an imaginary invariant plane that is perpendicular to the vertical.

Fast precession is rare with a real top because it is so much faster than the precession that depends on gravity. Much more energy is needed to send the top into fast precession. Unless the top is somehow given that energy during the initial stages (for example, if a fortuitous launching imparts the energy and the required initial motion), it will settle instead into slow precession.

With some care you can set a top spinning so that its body axis is vertical. (The technique of launching a top with two lengths of tape, which I shall describe below, is useful here.) If the spin rate is higher than some critical value, which varies according to the type of top, the top stays upright. At a spin rate lower than the critical value the top nutates between the vertical and some limiting circle corresponding to a larger angle to the vertical.

Both kinds of behavior can be seen in a real top with a spin rate that is initially above the critical value and a body axis that is initially vertical. The spin has two stages: first a sedate vertical spinning and then a frantic nutation. The transformation is brought about by the friction at the point where the top touches the floor. Although the initial spin rate is high enough for the vertical spinning, friction eventually pushes the spin rate below the critical value and the top begins to lean and to nutate and finally to topple.

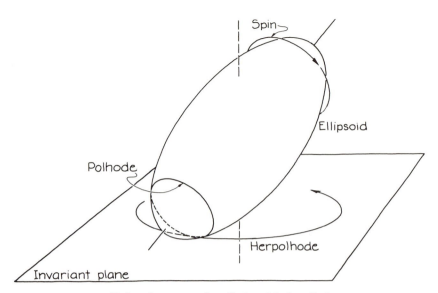

Motions that generate the polhode and the herpolhode

Can a top that is initially tilted ever nutate until it is vertical? In principle it cannot because it will always lack the energy to become precisely vertical. It can come close, however, if the upper limiting circle is quite small. Then the top can nutate upward in such a way that it passes close to the vertical before it nutates downward again. This situation will probably not arise often.

A keen observer of tops would not agree with much of what I have said. Many tops do rise to a vertical spin (such a top is called a sleeper), and so my statements are not entirely correct. A top rises not through a process of nutation but through a subtle interplay of the stem of the top and the friction between the point of the top (the end of the stem on which the top spins) and the floor. The stem of any real top is not the infinitesimal point I have been assuming so far. A better model would be a hemispherical stem.

Such a stem would tend to rotate on the floor because of the top's spin. Meanwhile the top is also trying to precess. Since I am now considering a more realistic stem instead of a stationary point of contact with the floor, I must visualize the precession as moving the top around its center of mass, which remains stationary. The contact point on the floor moves, circling below the center of mass, while the upper section of the top circles above the center of mass.

As a result the hemispherical stem is attempting to roll over the floor in two ways, one due to the spin of the top around its body axis and the other due to the precession driving the stem over the floor. The first way is much faster than the second, and so the stem slips. As it slips the friction between it and the floor points in a direction opposite to the direction of the slip, as is shown in the illustration on page 43. It is this friction that rotates the body axis of the top to

make the spin vertical. To calculate the torque I treat the pivot point as the center of mass. The lever arm is the distance between that point and the place where the stem touches the floor. The torque raising the top is equal to the multiplication of the lever arm and the friction.

Of course, the friction on the stem is also stealing energy from the top. The trick in designing a good sleeper is to shape the stem so that the top becomes vertical before it loses too much energy. Hence a good sleeper has a stem with a small radius of curvature. Tops with stems that have a more gradual curvature take longer to rise and may not reach the vertical before the spin declines below the critical value needed to sustain a vertical position. A whip top, which is lashed with a whip to add energy to the spin, has a stem with a very small radius of curvature; it rises to be a sleeper within seconds of being lashed.

Most of the commercially available tops are shaped like a pear. Some are spun by the fingertips. Others are wound with string and thrown to the floor. Some handcrafted tops are mounted in a wood holder until the string wrapped around their spindle is unwrapped with a brisk pull; then they fall spinning to the floor.

Although all these types are usually found in toyshops, they by no means exhaust the variety of tops that can be (or once could be) found around the world. A beautiful description of the many kinds of top that have been made can be found in the book by D. W. Gould cited in the bibliography for this issue [page 66]. Some tops have multiple faces for the purpose of gambling and telling fortunes. Some are flat disks pierced by a thin rod. Others are flat disks with a central bump on which they spin. (Several of them can be deployed at one time from a single holder.) The most remarkable feature of tops (apart from their rotational mechanics) is how they were developed in such a variety of shapes by people working almost independently around the world.

Dubois recently wrote me about the tops he and his students make from common items such as buttons, needles and machine screws. Often these unlikely objects need only a little fixing before they can serve as tops. A point may have to be ground or a slot may be required in a spindle so that a cord can be inserted. Sometimes several pieces are assembled into a top with glue. (Dubois says epoxy is best.)

Since friction plays a role in the behavior of tops, Dubois experiments on different surfaces. Paper, thin plastic sheets (such as food wrap) and bed sheets work well with some tops but not with others. Dubois has found that a linoleum floor is the only surface on which all his tops work adequately, although most of them work better on some other surface. You can keep track of where the point of a top travels by spinning it on carbon paper, soot-covered paper, ink-covered glass or some other surface on which the point can leave a mark.

Some of Dubois's tops can be spun by a quick snap of the fingers. He holds the upper stem of the top between his thumb and second finger, keeping both the point and his palm downward. Then he snaps those fingers, causing the top to spin outward on the table or the floor. If there is no upper stem, he holds the stem with his palm upward. Then when he

releases the top with a snap, the point is properly downward. Another launching procedure is to hold the stem between the forefinger of the left hand and the thumb of the right hand; when the two digits are pulled quickly in opposite directions, the top is spun.

Fast spins can be achieved with some tops by wrapping a cord or a strip of cloth around the stem or some other cylindrical part of the top. A typewriter ribbon works well. (The ink should of course be removed first.) You can slip one end through a notch in the stem and then wrap the ribbon around the stem several times. When you throw the top outward across the floor while holding the end of the ribbon, the top is made to spin rapidly.

This technique does not work well with small or fragile tops. With them Dubois prefers a double wrapping of ribbon around the stem. Both strips are started in the same way in the notch. As the top is turned in the hand the double layer of ribbon is wrapped around it six to 12 turns. Then instead of throwing the top to unwrap the ribbon, pull the two strips horizontally in opposite directions so that the top is twirled. With double-layer wraps of up to 14 revolutions Dubois can make an upholstery tack spin at about 15,000 revolutions per minute. With a finger snap the speed is about 8,000 r.p.m.

If the top has a smooth stem, as an upholstery tack does, the ribbon is difficult to wrap because it slips easily. Dubois coats the surface with a layer of

epoxy to provide more friction. The added friction also means that when the end of the ribbon is reached during the launching, the top is deflected by a final tug rather than slipping away freely.

Some of Dubois's tops sing as they spin, either because they scratch the surface on which they are spinning or because they generate turbulence and vibration in the air. The spin of a singing top must exceed 3,300 r.p.m. One of Dubois's faster tops sings two octaves above A (440 hertz). If the top is spinning on a membrane of some kind, the membrane acts as a sounding board to enhance the audibility of the sound. Do not stretch the membrane too tight or it may be punctured by the spinning top.

Spin and precession can be monitored with a repetitive-flash strobe light. The frequency of the light is varied in order to match either the spin or the precession. Some of the tiny tops Dubois has sent me turn at speeds of more than 100,000 r.p.m., which is possible because the top's moment of inertia around its body axis is small and the energy imparted at launching results in a high rotational speed. The commoner pear-shaped top has a much greater moment of inertia around its body axis, so that the launching energy results in a much lower rotational speed.

The most fascinating top I have ever encountered is the Tippe Top, which I described in this department for October, 1979. The top usually has a hemispherical bottom and a central stem. It is spun with the bottom downward, a logical position because that section is heavier than the stem. Soon after the top is released, however, it inverts so that it spins on the stem. The heavier hemispherical end is lifted, seemingly in defiance of gravity.

The cause of the odd inversion is friction with the surface on which the hemisphere is turning. Spinning the Tippe Top on a soot-covered pane of glass will give a picture of the path of the top. Such a picture, made and analyzed in 1960 by Frank F. Johnson of Hasbrouck Heights, N.J., appears in the bottom illustration on page 44. Johnson's analysis is cited in the bibliography for this issue [page 66].

What would a top do on a slightly inclined surface? Would its point of contact simply move around the way it does on a horizontal surface? Ledo Stefanini of the University of Bologna recently published a detailed answer to this question. Like other people studying the role of friction in the behavior of a top, he considered the stem to be hemispherical.

If the top is spinning without precession, it will move horizontally across the inclined plane in a straight line. If it is precessing, however, four things are possible, depending on the slope of the inclined plane and how much the top tilts from the vertical. One possible path

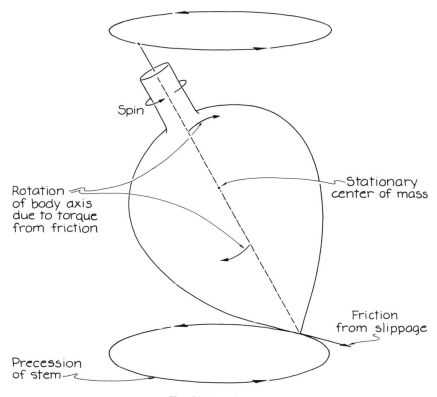

Spin

Rotation of body axis due to torque from friction

Stationary center of mass

Friction from slippage

Precession of stem

How friction raises a top

for the stem is a spiral along a line down the plane. The other patterns resemble the nutation patterns of an ideal top with a sharp stem that does not move across the floor. In some cases the stem traces out loops on the plane as it moves primarily in a horizontal direction. Or it may map out cusps. For another set of angles the stem first climbs the inclined plane and then descends while also moving horizontally across it.

You might like to try various combinations of angles on an inclined plane. Stefanini published a tracing he made from a top weaving across carbon paper he had tacked onto an inclined plane. You can make your own tracings to catalogue the possible motions.

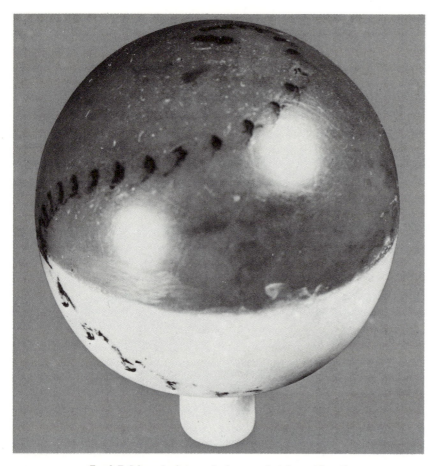

Frank F. Johnson's photograph of soot marks left on a Tippe Top

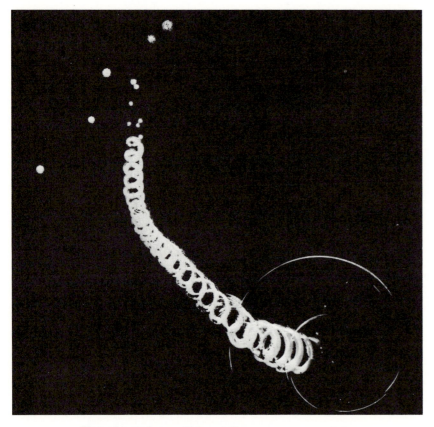

The track made by one of Johnson's Tippe tops on soot-covered glass

NOTES

Joseph L. Snider of Oberlin College and John Hart of Cincinnati, Ohio, corrected the illustration that shows how friction raises a top. The friction is certainly directed opposite to the slippage, but it should have been drawn in the opposite direction from that shown. Thus the friction points in the direction of the precession of the stem. To make sense of this orientation, consider the object in the preceding illustration to be slipping because its spin is larger than its precession. The bottom of the object slips from right to left in the illustration. Thus the friction it generates points from left to right, which is the direction of the precession. Snider also pointed out that the plane in that illustration is properly called the invariable plane, not the invariant plane.

Martin G. Olsson of the University of Wisconsin at Madison has corrected me on the occurrence of fast precession. "In the usual case fast precession is not 'rare' at all but is just ordinary nutation. To see this, imagine a top spinning rapidly in no gravity. Pushing such a top causes the tip to trace out a circle with the usual 'fast precession' frequency characteristic of a free top. Now if we turn on a weak gravity field the circular motion will hardly be affected. The average direction will now precess slowly and the net motion will consist of a superposition of circular motion and the uniform precession. The result is a trochoidal type of motion characteristic of nutation."

8

Wobbling

Delights of the "wobbler," a coin or a cylinder that precesses as it spins

Set a coin on its edge on a smooth surface and flick it with your finger. It will spin vertically but soon will start to wobble and tilt until, with a clatter that increases in frequency, it ends up flat on the surface. A cylindrical object such as a bottle can also be made to wobble on its base, but unless the wobble is so strong that it makes the object fall over it remains upright. A bit of experimenting reveals a lot about how such objects wobble and what determines whether a wobbler finishes flat or upright. My analysis is based on a study by Lorne A. Whitehead and Frank L. Curzon of the University of British Columbia that will appear in *American Journal of Physics*. Whitehead designed an apparatus for studying the phenomenon and initiated the mathematical analysis. I have also drawn on earlier work by Martin G. Olsson of the University of Wisconsin at Madison.

Wobbling has three noteworthy features. First, it is periodic. Second, the object rolls around on its supporting surface without slipping. Third, the cen-

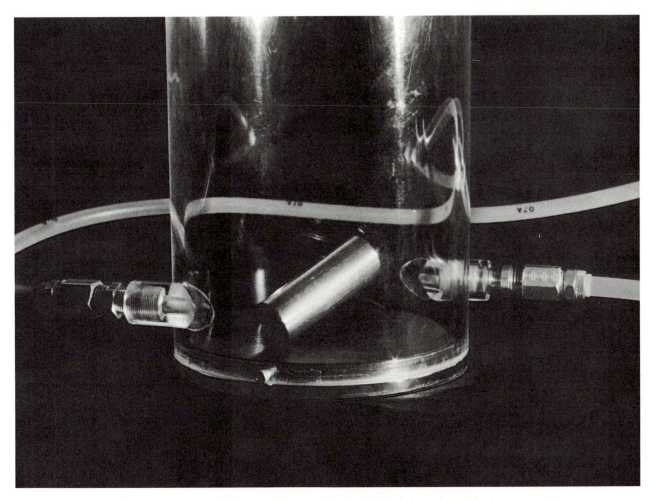

An aluminum cylinder wobbling in a chamber designed by Lorne A. Whitehead

ter of mass of the object gradually moves downward but does not shift much horizontally unless the object has been started into a clumsy spin.

Wobble lasts for as long as it does because two losses of energy are minimized. The lack of slippage diminishes the loss of energy to friction with the surface on which the object is spinning. The lack of vibration of the center of mass and of the support point eliminates an additional loss of energy.

The top illustration on this page shows the main physical characteristics of a wobbling object. The object is in contact with the supporting surface at a single point. An axis of symmetry extending along the length of the object is usually at an angle to the vertical. As the wobbler spins around this axis the axis turns around the vertical. Since the center of mass is stationary as the object rotates, the bottom of the wobbler rolls in a circle around the vertical axis passing through the center of mass. This is precession, the movement usually seen when a toy top has been set spinning.

The force that generates precession is developed at the point where the wobbler touches the surface on which it is turning. Although each part of the wobbler is pulled downward by gravity, the combined weight can be assumed to act through the center of mass. If the wobbler were not also spinning around its axis of symmetry, the pull on the center of mass would cause the object to fall over.

A force usually makes an object accelerate in the direction of the force, but a force acting on something that is spinning creates a torque that makes the object turn. The torque on a wobbler causes precession. If the rate of rotation is too low or too high for the object to roll without slipping, friction quickly changes the rate of spin until the slipping stops.

What the torque is actually doing is redirecting the angular momentum of the wobbler. This quantity is the product of the object's distribution of mass (its moment of inertia) and its rate of spin. The angular momentum can also be described as a vector that lies in an imaginary vertical plane containing the axis of symmetry around which the wobbler is spinning. Rotating about the vertical axis that passes through the center of mass, the vector maps out an imaginary cone. Hence the axis of symmetry also rotates about the vertical. Lacking gravity and the torque it generates, an object could be made to spin without precessing. A wobbler then would not wobble.

A wobbler's rate of precession depends in part on the angle between the axis of symmetry and the vertical. When Whitehead and Curzon analyzed this re-

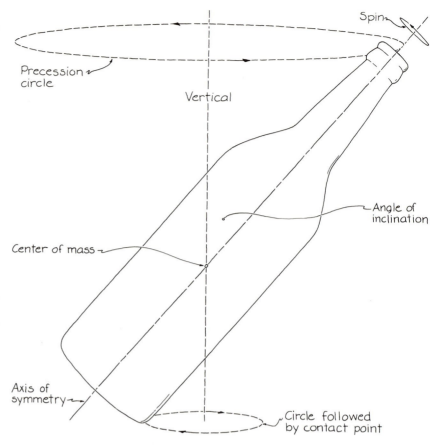

The physical characteristics of a wobbling object

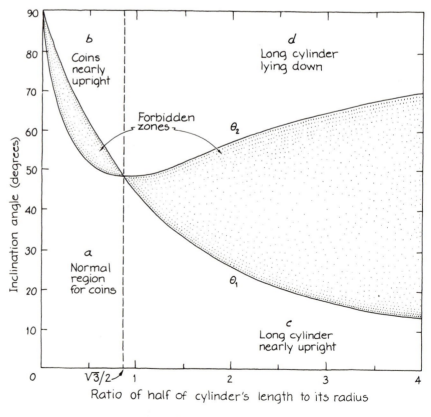

The conditions for steady precession

lation, a surprise emerged. In general a wobbler will precess steadily at any angle of inclination between approximately zero (with the axis of symmetry nearly vertical) and 90 degrees (with the axis nearly horizontal). The surprise was that the wobbler will not precess steadily in a certain zone that is determined by its shape. The forbidden zone is easy to determine if the object rolls around on a circular rim, as coins and bottles do, but is hard to determine if the shape is more irregular.

Whitehead and Curzon worked with a cylinder in their study of the requirements for steady precession. The range of inclination angles at which a cylindrical object cannot precess steadily depends on the shape of the cylinder. In particular the ratio of half of the cylinder's length to its radius is crucial. A coin is a cylinder in which the ratio is small; with beer cans and most other cylinders the ratio is large.

The results of the analysis are shown in the graph at the bottom of page 46. The ratios of half length to radius are on the horizontal axis and the inclination angles are on the vertical one. For example, the inclination angle of a coin is small when the coin is nearly flat and large when it is on or near its edge.

The cylinder precesses steadily when its axis of symmetry is outside the forbidden zone. That zone is marked by the angles θ_1 and θ_2. The angle is θ_1 when the cylinder's center of mass is directly above the support point, in which case gravity provides no torque for the precession. At θ_2 the angular-momentum vector is vertical and torque again does not cause precession.

Ordinarily these angles differ. They are the same only for a cylinder with a half-length-to-radius ratio equal to $\sqrt{3}/2$ (half of the square root of 3). A cylinder with a smaller ratio precesses more like a coin and would be recorded on the left side of the graph. Long cylinders such as cans precess in a way that is reflected on the right side.

A coin spun on a surface begins to wobble in a steady precession when the conditions are as indicated in section *a* of the graph. As it gradually loses energy to friction its angle of inclination decreases. Initially the rate of precession decreases too, but when the angle of inclination approaches zero, the precession rate begins to rise. You can detect the change with your ears as well as your eyes. As the angle of inclination decreases, the clatter of the coin first drops in frequency (reflecting the initial decrease in the rate of precession) and then rises.

If you are looking down on the coin, you can also monitor the spin rate. At first it is too high for you to be able to see the face of the coin clearly. As the coin approaches its final horizontal position

the spin rate diminishes and the face of the coin becomes clearer. The increasing clatter from the increasing precession rate contrasts sharply with the decreasing spin rate.

How can the precession rate of a wobbling coin increase while the coin is losing energy to friction? The coin has three types of energy. Two are rotational kinetic energies associated with the spin around the axis of symmetry and the precession around the vertical. The third type is gravitational potential energy. Every part of the coin except the point in contact with the table has potential energy, but the situation is easier to understand if the entire amount is associated with the center of mass.

The coin has to tilt increasingly as it wobbles because friction with the surface gradually removes the energy of the spin around the axis of symmetry. With a smaller rate of spin the axis of symmetry must move closer to the vertical. As a result the center of mass drops closer to the surface and the potential energy of the coin decreases. The loss

of energy (both kinetic and potential) forces the coin to precess faster, somewhat like a rubber ball that bounces more frequently as it loses energy.

In theory you could also make a coin spin in the conditions represented by section *b* of the graph, but it would be difficult. The coin begins to spin while it is nearly vertical. As it loses energy to friction its angle of inclination increases to 90 degrees. In principle the coin would come to rest balanced on its edge. This position is unlikely in practice because the coin is too sensitive to small perturbations. It would probably fall over and lie on the surface.

A cylinder with a ratio of half length to radius of $\sqrt{3}/2$ shows the greatest range of inclination angles at which it can precess smoothly. Longer cylinders such as bottles wobble in two different ways, which are represented by regions *c* and *d* on the graph. A cylinder set wobbling when it is nearly upright performs as shown in region *c*. You could start a bottle in this way by putting one hand on each side of it while it is inclined al-

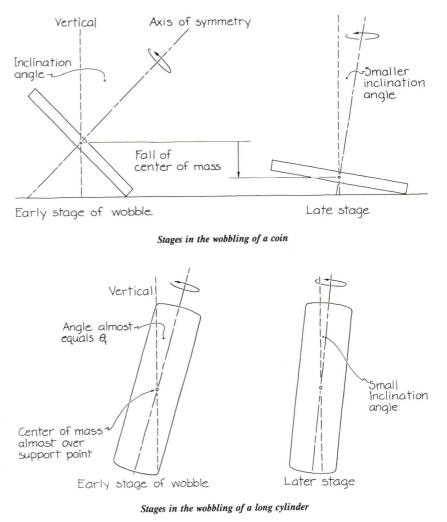

Stages in the wobbling of a coin

Stages in the wobbling of a long cylinder

most at the balancing angle of θ_1 and rapidly moving your hands in opposite directions.

The bottle rolls on its rim. It gradually speeds up, and the angle of inclination decreases as the bottle approaches the vertical. Since the bottle is tall, it can be affected by nonuniformities in the rim and the table. The wobble is usually unstable.

A cylinder wobbling under the conditions represented in region d begins with its axis of symmetry almost horizontal. The angle of inclination may be just slightly larger than θ_2. As with a coin the angle of inclination cannot be exactly θ_2 because the precession and spin would be infinitely fast. In losing energy the cylinder begins to lie down and its rate of precession decreases steadily. When it lies flat, it still has in principle some minimum rate of precession, but in fact it is quickly stopped by friction with the surface.

You can start a small cylinder in this type of motion with a snap of a finger. The cylinder precesses smoothly and gradually descends to the surface, where it rolls about more irregularly. Its clatter steadily decreases in frequency.

Whitehead realized early in his work that the wobble of a cylinder could be prolonged with jets of air. He rigged an apparatus in which jets of air were directed tangentially at a cylinder of aluminum eight centimeters long and three centimeters in diameter. Launched by hand at a large angle of inclination, the cylinder wobbled in a way represented by region d of the graph. Precession frequencies of up to 100 cycles per second could be achieved by adjusting the jets. Although the motion was stable, the cylinder had a tendency to drift across the table. Whitehead therefore set out to build a better apparatus.

The result is a container with a Plexiglas base 15 centimeters in diameter. Whitehead machined the base so that it is concave; its radius of curvature is 50 centimeters. The top surface is polished to eliminate rough features that would interfere with a stable wobble. A layer of rubber an inch thick is glued to the surface to provide better resistance to the air. The concave surface keeps the wobbler from straying from the center.

Air jets from a standard laboratory supply of air are directed into the container near the base. They maintain the energy of the wobbler, overcoming its losses to friction. The air goes out of the container through the top, which is a flat piece of Plexiglas with holes in it.

The main problem faced by Whitehead was how to freeze both the precession and the spin of the wobbler. Usually the two proceed at different rates. The solution was to deploy a wobbler in which the spin rate is an integral multiple of the precession rate. If a stroboscope is then set to flash at the precession rate, both the precession and the spin are frozen.

Whitehead chose as a wobbler a carefully machined cylinder of aluminum with a half-length-to-radius ratio of 2.71. When such a cylinder wobbles at an angle of inclination of 64 degrees, its

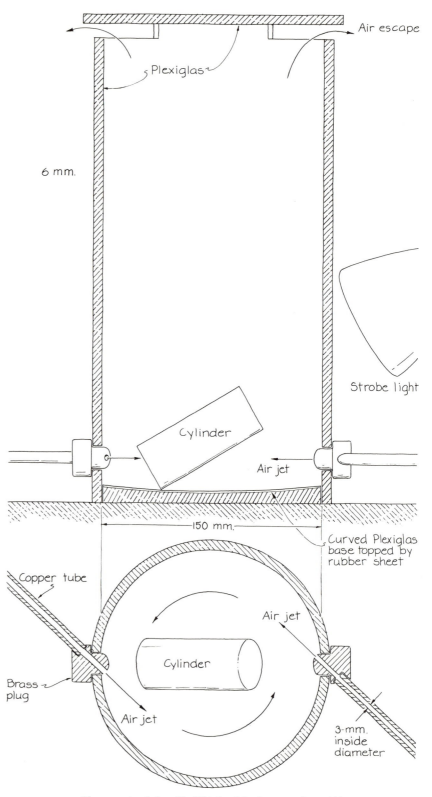

The apparatus designed by Whitehead for demonstrating wobble

spin rate should be twice as high as its precession rate. That angle is close to θ_2 for the cylinder. By adjusting the rate of flow in the air jets he could achieve fine control of the angle of inclination and thereby of the precession rate. The stroboscope was operated at a frequency of 50 hertz.

Whitehead had to remachine the cylinder a bit to achieve the desired ratio of 2 in its precession and spin rates. A precision of about .1 millimeter was necessary. When the cylinder was right, he set it in motion by hand to wobble under the conditions of region d. It quickly developed a stable wobble near the desired angle of inclination. Whitehead adjusted the flow of air to maintain the angle. Under these conditions the wobble could be maintained for three or four days. A cylinder of this kind could be kept wobbling longer if the concave base of the apparatus were made of a material harder than Plexiglas. To keep wobbling smooth on Plexiglas calls for

an occasional repolishing of the surface.

Olsson has studied the wobble of a coin, analyzing the motion with methods that could also be applied to a child's top. The two objects are actually quite similar. When Olsson sets a coin spinning on its edge, the spin is at first stable against perturbations from the table and from gravity. Friction gradually slows the spin, which soon drops below a critical value determined by the coin's mass, radius and moment of inertia. Then the coin begins to wobble.

Olsson also demonstrates wobbling in a large-scale way, spinning an aluminum disk that is an inch thick and about as big as a manhole cover. It wobbles with quite a noise, and toward the end, when the disk is almost flat, the racket is impressive. In a classroom where the seats are on a sloping floor the students can both watch the spin rate and hear the precession rate. The contrast between the decrease of the one and the increase of the other is mesmerizing.

NOTES

The original publication of this article had several errors that have been corrected here. The most important errors were in the graph, where the labels "a" and "b" had been interchanged and the label of half the square root of three failed to point to the dashed line.

The layer of rubber employed in protecting the surface on which the wobbler spun was 1/16 inch thick (not one inch as stated). Such a protective rubber layer is not essential to the experiment. However, if the wobbler spins directly on the Plexiglas (or even harder materials), you must occasionally repolish the surface to avoid wear marks that would interfere with the wobbling. Additional information about wobbling can be found in the article by Whitehead and Curzon, which has now been published and is listed in the bibliography.

9 Boomerangs

How to make them and also how they fly; a few notes on Frisbees

A boomerang is surely one of the oddest devices ever to serve as a weapon or a plaything. It was apparently invented by accident, notably in Australia by the native people but also independently in many other places. If you throw an ordinary stick, it falls to the ground not far away, but a boomerang can travel as much as 200 meters (round trip) and can be aimed so skillfully by an expert thrower that game or an enemy can be hit. The boomerang probably originated as a weapon designed for accurate straight flight, but most people find the returning boomerang, which was mainly just a plaything for the Australians, more interesting. Ironically the straight-flying boomerang is probably more complicated aerodynamically than the returning boomerang. As ancient as both devices are, an amateur investigator can still do a great deal to advance the understanding of their flight.

Although good returning boomerangs are occasionally available in sporting-goods stores, most commercial returning boomerangs are mass-produced and fly poorly. Indeed, many of them do not even return. The plastic boomerang made by Wham-O is one of the best types available in toy stores. Many other types, all of them excellent, are available from Ruhe-Rangs, Box 7324, Benjamin Franklin Station, Washington, D.C. 20044, thanks to Benjamin Ruhe, a boomerang enthusiast formerly with the Smithsonian Institution. During his service there Ruhe helped to organize the Annual Smithsonian Open Boomerang Tournament. The tournament, held in late spring or early summer, is great fun for the 100 or so contestants who enter. This year it is scheduled for June 9 on the mall in Washington.

If you would like to experiment with boomerangs, you need to be able to construct your own. Only then can you make the variations necessary to determine what factors influence a boomerang's flight. The best material from which to cut the basic stock of a boomerang is Baltic birch in a marine or aircraft plywood between 1/4 and 3/8 inch thick, with five or more laminations. This type of plywood is resistant to wear and water; it is also dense, which makes for a boomerang that is heavy for its size.

Cut a cardboard pattern for a boomerang of whatever shape you want. An example is the returning boomerang designed by Herb Smith and shown in the illustration at the left. (If you are left-handed, you will have to make a left-handed boomerang, which is the mirror image of the one drawn.) Place the pattern on the plywood and mark off the outline of the boomerang in pencil.

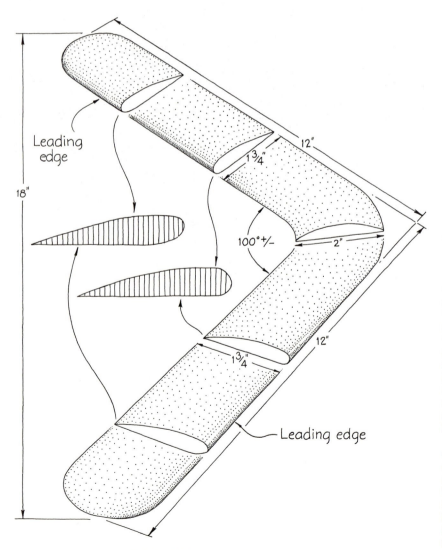

Herb Smith's "Gem" design for a boomerang

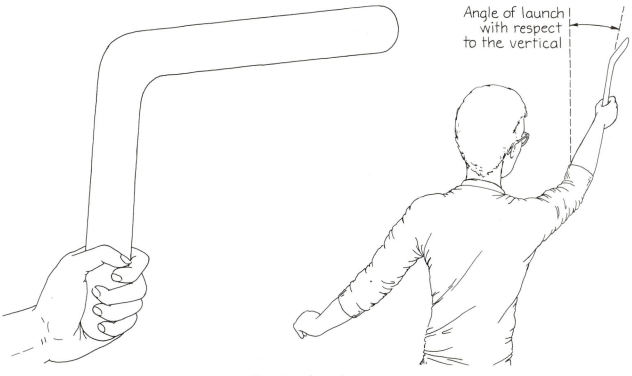

Preparing to throw a boomerang

With a coping saw or a handsaw cut out the boomerang blank. The edges and the top must now be shaped in the general form shown in the illustration. (Little is done to the bottom other than putting a slope on what will be the leading edge.) The leading edge must be blunt, whereas the trailing edge must be sharper, with the top surface sloping down to meet the unaltered, flat bottom surface. The arms must have the cross-sectional shape of a classic airfoil, because they must provide lift much as the classic airfoil does.

Clamp the blank in a vise and cut and shape the edges and the top with a rasp that has a curved surface. Smooth out the grooves left by the rasp and finish shaping the wood by rubbing the surfaces with a piece of coarse sandpaper wrapped around a piece of soft wood. Before you finish off the surface with a finer sandpaper you should make a test flight with the boomerang so that you can tune it by more shaping with the rasp or with coarse sandpaper. Tuning the boomerang means that you throw it, cut or sand it more and then throw it again until it flies the way you want it to.

A boomerang cannot be thrown well in a strong wind. If there is a light wind, face toward it, turn 45 degrees to the right and throw the boomerang in that direction. Hold the boomerang vertically by the tip of one of the arms (which one usually does not matter) with the flat side away from you. Reach behind your head with the boomerang and then throw it toward the horizon, snapping it forward when your arm is fully stretched forward. Do not try to throw hard at first. It is the snap that counts, not the strength of the throw. The boomerang stays up in the air because of the spin the snap imparts to it.

The proper orientation of the boomerang (the plane in which it will be spinning) will vary according to the wind conditions and the type of boomerang. To achieve a good flight you might have to throw the boomerang with its plane nearly vertical. Under other circumstances you will have to shift the plane (tilting the top of the boomerang away from you) by as much as 45 degrees. The greater the tilt of the plane in which the boomerang spins is, the more upward lift the boomerang will initially have. If you give it too much lift, it climbs too fast and then plummets to the ground with such force that it may break.

In a good flight a returning boomerang travels horizontally around an imaginary sphere. On returning it probably will hover or even loop a bit before it drops to the ground at your feet. If you are lucky, it might make one or two additional circles (smaller than the first circle) before it falls. Although you launch the boomerang with its spin plane almost vertical, it probably will return with the plane nearly horizontal. I shall explain below why the plane must turn over in this way if the boomerang is to complete the trip.

If your boomerang consistently lands to your right in a light wind, try throw-ing it a bit more to the left of the wind. Similarly, if it is landing to your left, try throwing it more to the right of the wind. If it lands behind you, try throwing it with less force. If that does not work, throw it somewhat above the horizon with the spin plane tilted less from the vertical. If the day is windless and you are not getting a full return, tilt the spin plane farther from the vertical in order to gain more lift during the flight.

Be careful not to injure people or damage things with your boomerang. It can be quite a weapon. Throw it only in a large, open space. If people are present, make sure they know what you are doing so that they can be prepared to dodge.

The successful tuning of a boomerang involves both experience and luck. In general if you make the top surface more curved, the boomerang will have more lift, which means that it will return in a tighter circle. Flattening the top surface or curving the bottom surface will give the boomerang less lift because the cross-sectional shape of the arms is then less like a classic airfoil. If the spin decreases too fast, so that the boomerang falls to the ground in mid-flight, the reason may be that excessive air drag on the arms is robbing them of their spin. Some surface roughening might be beneficial to the flight, but any large grooves left by the rasp will surely create additional air drag that will shorten the flight time.

Instead of shaping the arms carefully you might prefer to twist them so that during a flight the leading edge on each

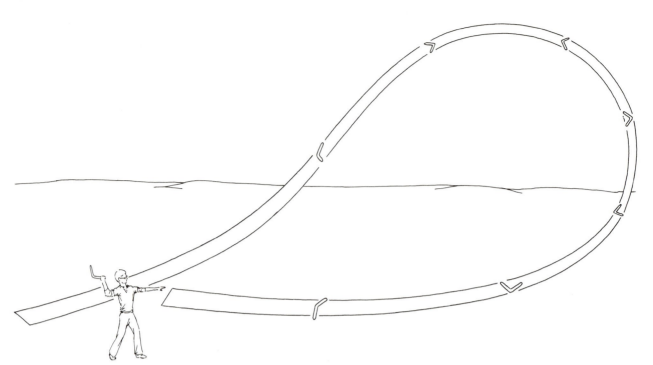

A typical flight path

arm is tilted to deflect the passing air to the right, giving the boomerang a lift to the left. This type of lift is easy to visualize. When you put your hand out of a window of a moving car and turn it through various angles, you can feel the lift. To twist a boomerang heat it gradually in an oven at 400 degrees Fahrenheit and then (with gloves on, of course) carefully twist the arms until the wood is cool. If you twist too much, heat the boomerang again and twist the arms back a bit.

Once you have a properly flying boomerang you might want to finish it off with a cellulose covering and some decorative designs. Smith's excellent booklet, which is listed in the bibliography for this issue [*page 67*], explains how to do this kind of finishing and also gives a number of boomerang designs.

If your boomerang breaks, do not throw away the pieces. Glue them together with epoxy, clamp them until they dry and then file and sand the surface back into the desired shape. Although the boomerang will not be as strong as it was before, its flight path might be altered in an interesting way by the small change in its distribution of mass resulting from the break.

A boomerang does not have to be limited to two arms. Indeed, one of the easiest boomerangs to build is a four-blade design consisting of two rulers crossed and fastened at the center. The right kind of ruler has a curved top surface and a fairly flat bottom surface. You can attach two such crossed rulers either by wrapping a strong rubber band around them or by putting a bolt and nut through the center hole they usually

have. Throw this type of boomerang the same way that you would a two-armed one. Take care to avoid being cut by sharp edges, and never use rulers that have metal edges.

A simple cross boomerang can be fashioned from a cardboard square about five inches on a side. Cut out a boomerang with three or four blades and twist them slightly so that the boomerang is not quite all in the same plane. Adding weight to the arms of a boomerang increases its range. With the cardboard boomerang it is easy to add weight by attaching paper clips to the end of the arms. This boomerang can be demonstrated in a classroom. If it has too much range for a classroom, you

can decrease the range by increasing the twist on the arms so that the boomerang is less in the same plane or by bending the arms along a center line through their length. With the latter technique the arms have an exaggerated airfoil shape: at least one side is sharply convex but the other is not. As usual, throw the boomerang with the convex side toward you. By changing the arms from being almost flat to being more like an airfoil, you increase the lift on the boomerang; its path will be a tighter circle.

When you begin to throw your wood boomerang well, you might be tempted to catch it. The result may be a sharp blow to your fingers. If you are determined to catch a boomerang, hold your

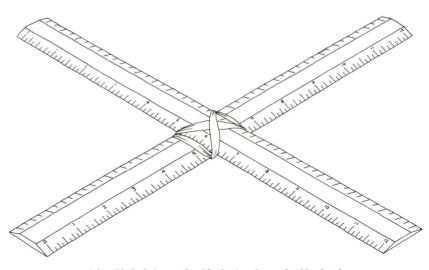

A four-blade design made with plastic rulers and rubber bands

hands flat and slap them together to trap the boomerang as it hovers above the ground, still spinning, in the last stage of its flight. Keep your fingers away from the turning blades.

The explanation of the return of a boomerang lies primarily in the cross-sectional shape of the arms and the fact that the boomerang spins. Without these two features a boomerang would behave like any other thrown stick. The cross-sectional shape gives the boomerang aerodynamic lift similar to the lift generated by some airplane wings. The spinning gives the boomerang stability. Through a bit of fortunate rotational mechanics the spinning also causes the axis about which the boomerang spins to rotate in much the same way that the spin axis of a top rotates about the vertical. The lift and the stability keep the boomerang up, and the rotation of the spin axis brings it back to the thrower.

Aerodynamic lift can be explained with a simple model of a classic airplane wing similar to the one I described in this department for February, 1978, to explain the lift of a kite. The classic airfoil has a flat bottom, a blunt front, a sharp rear and a convex top. Air passes around a wing faster along the top of the wing than along the bottom. The reason can be seen by visualizing the passing air as being of two kinds. One kind flows around the wing with no rotation in the stream and with the same speed on the top and the bottom of the wing. The other is a circulation cell: it flows to the rear over the top of the wing and to the front over the bottom. Such a circulation is created by a real wing because the air's viscosity and its adhesion to the surface of the wing force it into this pattern as it flows to the rear off the curved top surface.

In the superposition of the two idealized airstreams the two velocities add above the wing and subtract below it, with the result that the real air speed is greater above the wing than it is below it. The difference is important to the lift because the air pressure in the stream is inversely related to the speed of the stream. Hence the air pressure is less above the wing than below it, and the wing gets a push upward. (A real airplane wing can have a more complicated airflow pattern than this simple model implies. Moreover, when an airplane is traveling at high speed, some of the lift may come from the impact of the passing air on the underside of a wing that is inclined slightly upward in order to deflect the air downward.)

If the classic airfoil is inclined to the airstream in such a way that the airstream is more incident on the curved top side, the lift is of course less. Such an arrangement is termed a negative angle of attack. In a simple model the reduction of lift is due to the downward push

the incident stream exerts on the top surface. One might also argue that lift is partially lost because the tendency for the air to circle about the wing is lessened and the speed of the air on the top side of the airfoil differs less from the speed on the bottom side.

Conversely, if the airfoil is inclined so that the airstream is incident somewhat more on the flat bottom side than on the top, a situation that would be called a positive angle of attack, the lift increases because of the upward push from the airstream on the bottom side. The air drag also increases. If the angle is too large, the disadvantages of increased air drag outweigh the advantages of lift. The attack angle of the arms of the boomerang as they turn through the air is important to its flight.

Boomerang arms can have a variety of cross-sectional shapes, but most of them are similar in cross section to the classic airfoil. Usually this shape includes a blunt edge that turns into the air as the boomerang spins and a sharper edge that trails during the turn. One side is usually flat and the other convex. Variations on this basic form are numerous, however, and little systematic work seems to have been done on determining which shapes are best aerodynamically. Some boomerangs are actually flat on both sides but with their arms twisted so that the air is deflected as the arms turn through the wind.

The lift on a boomerang differs in a major way from the lift on the classic

airplane wing. In the first stage of a flight the boomerang's "lift" is mostly horizontal, with only enough upward force to balance the weight of the device. Since the boomerang is spinning mostly about a horizontal axis, the curved sides of the arms spin in a plane that is almost vertical and the lift is almost horizontal. For the sake of simplicity in what follows I shall ignore the weight of the boomerang. I shall also assume a boomerang that is thrown outward by a right-handed thrower so that the plane of spinning is initially exactly vertical. The lift will be to the thrower's left, so that the boomerang begins to move to the left as it continues to spin in the vertical plane.

If this were the entire story, the boomerang would never come back. To see why it turns around and returns you must understand what else the lift does to the boomerang. In particular it is necessary to know how the torque due to the lift on the boomerang causes a precession of the spin plane.

Imagine that one of the boomerang's arms has spun to its highest possible position and the other arm is almost in its lowest possible position. (I am discussing the basic banana-shaped boomerang.) The upper arm is turning in the same direction in which the center of the boomerang is moving, whereas the lower arm is moving opposite to the motion of the center. The air passing the upper arm is moving faster (in relation to the arm) than the air passing the lower arm.

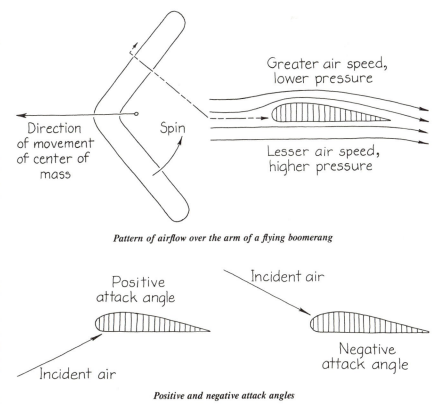

Pattern of airflow over the arm of a flying boomerang

Positive and negative attack angles

Therefore more lift is generated on the upper arm than on the lower one. The part of the boomerang higher in the spin will always experience a greater lift and hence a greater push to the side than the part lower in the spin.

My first thought was that the difference in horizontal lift (more lift on the upper arm than on the lower one) would cause the spin plane of the boomerang to tilt, thereby angling the lift downward (a disastrous effect). What actually happens, however, is that the difference in lift causes a rotation of the plane about a vertical axis. It is this rotation of the spin plane, commonly called precession, that brings the boomerang back.

To understand what causes the rotation you must examine the torque created by the lift. Take the center of the boomerang as the axis about which it is spinning. (Actually the center of mass around which a two-armed boomerang spins is likely to be well off center, but that does not alter the outcome of the argument.) Take the average lift on the upper arm as being directed horizontally outward from the center of the arm. Similarly, take the average lift on the lower arm as being also directed horizontally outward from the center of the arm. The torque created by one of these lifts, as measured from the center of the boomerang, is the product of the lift and the distance to where the lift is applied, that is, half the length of an arm. Since the upper arm has the greater lift, it also has the greater torque.

If the boomerang were not spinning, this difference in torques would merely make the plane of the boomerang tilt over. Since the upper arm has the greater torque, the plane would tilt counterclockwise as seen by the person who has just thrown the boomerang. The fact that the boomerang is spinning, however, makes a big difference, because it then has angular momentum and the tendency to tilt the spin plane results in a rotation of the spin plane about the vertical axis.

Angular momentum is the product of the boomerang's rate of spin and a function involving the mass and the mass distribution of the device. For an example in another setting imagine yourself attempting to turn a merry-go-round holding several children. The force you apply tangent to the rim multiplied by the radius of the merry-go-round is the torque you are supplying. When you begin, the torque causes an angular acceleration of the merry-go-round; the spin increases from zero to some final value. How would you arrange the children in order to achieve a given angular acceleration with the least force? Intuitively you would place them near the center. Their mass is the same, of course, but their mass distribution with respect to the center of rotation is different. When

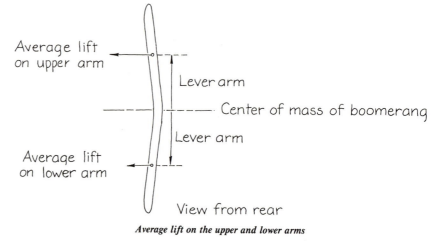

Average lift on the upper and lower arms

the mass is nearer the center, the merry-go-round is easier to turn. The mass and its distribution are taken into account by the function known as the moment of inertia. The greater the mass or the farther from the center it is placed, the greater the moment of inertia and the greater the force you will have to supply in order to achieve a given angular acceleration.

Once the merry-go-round is spinning and you are no longer pushing on the rim, the apparatus has a certain angular momentum because it has spin and a moment of inertia. Angular momentum is usually represented by a vector pointing perpendicularly to the plane in which the object is spinning. Here the vector would be vertical. The direction (up or down) is chosen by convention as being the direction of the thumb on the right hand when the hand is held in a hitchhiker's pose with the fingers curled in the direction of the spin of the object.

The only way you could change the size or the direction of such a vector would be to apply another torque to the object. With a merry-go-round you could push on the rim again. (A convention for choosing how to draw a vector representing the change in angular momentum involves pointing the index finger of the right hand from the center of the rotation toward the place where the force is applied and pointing the middle finger in the direction of the applied force. If you make your thumb on that hand perpendicular to both fingers, it automatically points in the direction of the change in the angular momentum. The new angular-momentum vector is the combination of the old one and the vector representing the change.) With a merry-go-round that you had resumed pushing tangent to the rim the new vector would still be vertical but would be larger or smaller depending on whether your aim was to make the merry-go-round turn faster or slower.

A boomerang that is spinning has two

torques acting on the arms, one created by the average lift on the upper arm and one created by the average lift on the lower arm. Since the lift on the upper arm is greater, it determines what happens to the angular momentum, and so I shall ignore the lift on the lower arm. (The argument would not change even if I included the smaller lift.) Imagine that the boomerang is receding from you just after you have thrown it with your right hand. It is spinning in a vertical plane and has an angular-momentum vector pointing to your left. The average lift on the upper arm creates a torque that will change the direction of the vector as the boomerang continues to fly away.

To determine how the vector changes use your right hand, orienting the fingers and the thumb properly. With your index finger pointing from the center of the boomerang to the center of the upper arm and your middle finger pointing to your left in order to be in the direction of the lift on that arm, your outstretched thumb points toward you. Therefore the vector representing the change in angular momentum points toward you. Mentally combining the change vector and the original vector is best done from an overhead point of view. The change vector is perpendicular to the original one and gives a new vector rotated from the old one toward you. The size of the angular momentum is unaltered because the change vector is perpendicular to the old one. Only the direction of the angular momentum is changed, and it is rotated about a vertical axis to point more toward you.

This type of rotation of an angular-momentum vector is precession; it is seen when the axis of a top precesses about the vertical. Another common example of precession is seen in the turning of a motorcycle. The wheels of a motorcycle spin fast enough and have moments of inertia sufficiently large to make their angular momentum large. To turn the motorcycle you cannot just

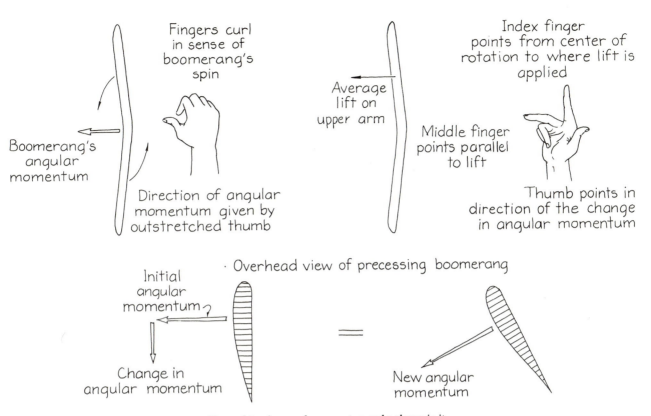

How to determine angular momentum and a change in it

turn the handlebars, as you would when riding a bicycle. Instead you make the motorcycle lean in the direction of the turn. The torques then experienced by the motorcycle cause the angular-momentum vectors of the wheels to precess, turning the motorcycle as a whole.

During the precession of the spin plane of a boomerang the boomerang continues to travel along a path with a certain speed but is continuously deflected by the horizontal lift it experiences. The resulting path approximates a large circle. In a successful boomerang flight the spin plane will precess at the same rate at which the device circles in

its path. Its angle of attack remains somewhat positive. This match is necessary in order to keep the arms at the proper attack angle.

Suppose the boomerang precesses too slowly. Then as it travels along its circular path its spin plane rotates about a vertical axis at a rate lower than the rate at which the boomerang as a whole travels along its path. When the spin plane lags behind, the attack angle becomes increasingly negative and the boomerang loses lift.

If the spin plane precesses too quickly, it turns about a vertical axis faster than the boomerang as a whole travels along

the large circular path. As a result the attack angle becomes increasingly positive until the spin plane is perpendicular to the oncoming airstream. The air drag would surely ruin the flight by then.

The match between the precessional rate and the rate at which the boomerang travels along the large circular path is not critical and is in fact somewhat automatic, since both rates depend on the lift. Throw your boomerang, sand down and reshape the arms and throw it again until you come near the match and the boomerang returns. I know of no sure way to remedy a persistently unsuccessful boomerang.

The circular path of the boomerang is independent of the speed with which you throw it. Only the moment of inertia and the cross-sectional shape of the boomerang determine the radius of the path. With a given boomerang you will therefore achieve the same large circle (for the same throw of the boomerang in the vertical plane I have been assuming) regardless of how hard you throw the device (provided, of course, you throw it hard enough so that it has sufficient speed to complete its journey). If you want to change the size of the circle, you must ordinarily choose a different boomerang with a different moment of inertia or cross-sectional shape. Next month, however, I shall explain how you can also add ballast to the arms in order to increase their moment of iner-

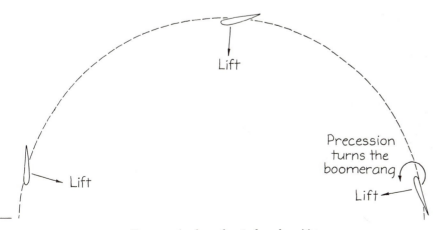

How precession keeps the attack angle positive

tia. This technique is used by boomerang throwers intent on breaking distance records.

A Frisbee flies in a quite similar way. It has a curved top surface and is launched with a flick of the wrist to give it spin. The Frisbee gains lift by virtue of the impact of the air or by the difference in air speed across the top and bottom surfaces. A Frisbee properly thrown in an almost vertical plane will return to the thrower the way a boomerang does.

Usually, however, a Frisbee is launched to curve slightly to another person, so that the thrower orients the spin plane to provide just enough horizontal lift to achieve the curve.

Both a boomerang and a Frisbee can be skipped across the ground without destroying the flight. Imagine a Frisbee skimming just above the ground with its leading edge tipped slightly downward. That edge then strikes the ground. The force from the ground at the contact point puts a torque on the Frisbee and changes the angular momentum, but because the change vector is almost perpendicular to the original angular-momentum vector the new angular-momentum vector is just a rotation of the original one. The angular momentum does not change appreciably in size, only in direction. Therefore the spin of the Frisbee is not much slowed; the device is merely reoriented and then goes sailing off in a new direction.

More on Boomerangs

Their aerodynamics and connection with the dimpled golf ball

Last month I described the basic mechanism that makes a boomerang return to the person who throws it. My account left some of the stranger aspects of the design and flight of boomerangs unexplained. This month I shall first deal with these aspects and then explain how you can experiment with the boomerangs you have made according to the techniques I described last month.

One interesting feature of certain older boomerangs is that they have a rough surface. Some of the ancient boomerangs found in Australia have this characteristic, as if the person who made the boomerang thought a rough surface would make it go farther. Although the idea seems implausible, it might be correct. One's initial reaction is that roughening would increase the friction between the surface and the passing air, so that the boomerang would lose forward velocity sooner and fall to the ground earlier. If a rough surface is desirable, the reason certainly has nothing to do with friction. Something else must be going on.

Many familiar objects have surfaces that have been made rough to increase their flight time. The dimpled golf ball is the best-known example. In the early days of golf all the balls were smooth, as intuition would suggest. Eventually someone noted that a ball scarred and pocked by use seemed to travel farther than a smooth one when both were struck in essentially the same way. Thereafter balls were pitted on purpose. Ascher H. Shapiro once did experiments comparing a dimpled golf ball with a smooth one. He found that the dimpled ball traveled more than four times farther than the smooth one, notwithstanding the difference in air friction.

Why would dimples aid an object in its flight? The answer lies in the behavior of the air flowing around a dimpled golf ball or a roughened boomerang. For the sake of simplicity first consider the flow pattern around a classic airfoil. Most of the air passing the airfoil is unaffected by the viscosity of the air, that

is, by the friction between two layers of air or between a layer of air and the surface of the airfoil. Next to the surface, however, is a layer of air called the boundary layer that is affected by viscosity in an important way. The air closest to the surface does not move at all, being held in place by the friction between it and the surface. The next layer of air has a low velocity. As one considers more of these imaginary thin layers of air at greater distances from the surface one finally reaches layers moving with a speed that is unaffected by friction with the surface of the airfoil. The thickness of the boundary layer is defined as the distance from the surface to the level where the friction can be neglected. The clue to the behavior of dimpled golf balls and roughened boomerangs lies in the movement of the air in the boundary layer.

The air pressure around a classic airfoil is high in front of the airfoil and behind it, relatively low over the top of it and relatively high below it. Last month I explained that the pressure difference between the top and the bottom is due to the Bernoulli principle. That pressure difference gives lift to the airfoil. This month I shall examine how the air moves from the front to the rear over the top of the airfoil.

Consider two small parcels of air, one parcel traveling inside the boundary layer and the other one just outside it. When they approach the airfoil, they are slowed to a stop by the high-pressure area just in front of the airfoil. When they reach the top of the airfoil (because it is moving), they must accelerate as they are pushed by the pressure difference between the front and the top. The parcel in the boundary layer must act against the viscous forces extending outward from the surface of the airfoil, however, and so it does not accelerate as much. When the two parcels pass over the top of the airfoil, the one outside the boundary layer is moving faster.

As the parcels then move toward the rear of the airfoil they encounter an increase in pressure and so slow down.

The parcel outside the boundary layer will be brought to a stop by the time it reaches the rear of the airfoil and then will be accelerated farther rearward by the high pressure there until it regains the speed it had before it began to pass the airfoil. The parcel inside the boundary layer begins the trip to the rear with a lower speed, so that it may stop before it reaches the rear of the airfoil. If it does stop, the high pressure at the rear may even push the parcel back toward the top of the airfoil, forcing air emerging from below the airfoil up to replace it. The air of the boundary layer will be pushed away from the airfoil by this influx, an effect that is termed separation and is an important factor in the drag on the airfoil.

Part of the drag comes from the friction between the passing air and the surface of the airfoil. One can call this contribution skin-friction drag. Another contribution comes from the difference between the average pressure on the front and that on the rear of the airfoil. If the average pressures are about equal, this contribution is small and one need consider only the skin-friction drag. In some cases, however, the pressure difference is even larger than the skin-friction drag. Such a situation may arise when the boundary layer separates from the surface, leaving a relatively wide wake of turbulent air replacing what initially was high pressure. The pressure of the turbulence is intermediate between the low pressure on top of the airfoil and the high pressure in front of it. Thus the pressure difference between the front and the rear may be large, creating a strong drag on the airfoil. The earlier the separation develops in the passage of air over the top of the airfoil, the stronger the drag from the pressure difference will be. With an earlier separation the pressure in the wake is lower and the wake is wider; both effects reduce the average pressure on the rear of the airfoil.

A smooth golf ball is a blunt object from which the boundary layer separates early, perhaps even before the air has gone halfway to the rear. A normal golf ball is dimpled in order to delay the separation. (Streamlining would accomplish the same result, but a golf ball shaped like an airfoil would be unlikely to roll well on the green.) At first the effect of the dimples seems counterintuitive because the dimpling surely also increases the skin-friction drag. The reduction in the pressure-difference drag is so large, however, that the overall drag is decreased and the golf ball travels considerably farther.

The dimples are designed to create a turbulent boundary layer that will rapidly mix the air in the boundary layer and the air outside the layer. As a result the air in the boundary layer does not have a chance to slow down with respect to the outside air because it is always receiving momentum from the outside air. When boundary-layer air passes over the top of the ball and heads toward the rear, it does not slow to a stop before it reaches the rear and the high pressure at the rear does not push it away from the surface. The wake is therefore relatively narrow, and the pressure difference between the front and the rear of the ball is not as large as it would have been if the ball had been smooth.

For the past several years a golf ball with a new dimple design has been offered by Uniroyal. The regular arrangement of circular dimples is replaced with a random arrangement of hexagonal dimples. Based on the research of John Nicolaides of the University of Notre Dame, the ball reportedly travels an average of six yards farther than the conventional ball. I am not sure why, but the cause is likely to be the creation of a better turbulent boundary layer.

Recently I saw a demonstration of the effect of dimpled golf balls at the Ontario Science Centre in Toronto. One smooth ball was attached to a vertical rod hinged at the top so that the ball could move when an airstream was directed at it. A dimpled ball was mounted in the same way. When the airstream was turned on, the ball with the lesser drag would be displaced less and its rod would swing less from the vertical. When I turned on the air, I found that the dimpled ball was displaced the least. The dimpled ball has less drag.

Should a boomerang be roughened or dimpled? Maybe. One must bear in mind the likelihood of separation on an airfoil traveling at moderate speed, as a boomerang does. When investigators study the flow of a fluid past objects, they must be able to compare the likelihood of separation and turbulence even though they are using different fluids (having different densities and viscosities) and different objects (having different sizes). To make the comparison a dimensionless number called the Reyn-

olds number is computed. Named for Osborne Reynolds, who contributed greatly to the analysis of fluid dynamics in the 19th century, the number is calculated from the density and speed of the fluid multiplied by a typical length of the object and divided by the viscosity of the fluid. When the Reynolds number is high, separation and turbulence are likely, the boundary layer is turbulent and the flow is said to be only slightly viscous. The fluid may actually have a

high viscosity, but under the circumstances of the high Reynolds number the flow is almost as it would be if there were no viscosity. A given pressure difference propelling a parcel of the fluid will create an acceleration of the parcel that depends on the density of the fluid rather than on the viscosity.

When the Reynolds number is low, separation and turbulence are unlikely, the boundary layer is laminar and the flow is said to be viscous. Again the be-

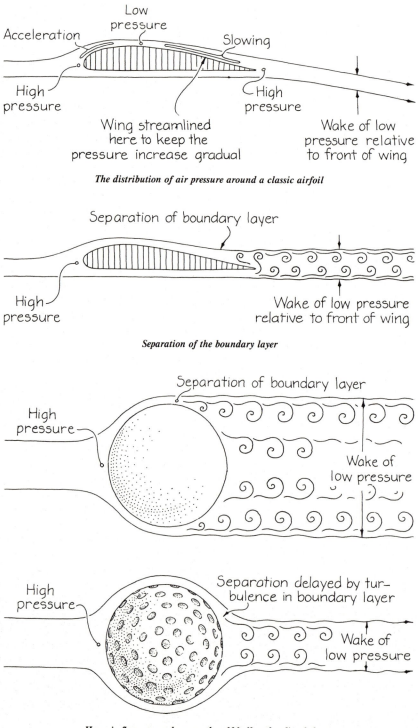

The distribution of air pressure around a classic airfoil

Separation of the boundary layer

How air flows around a smooth golf ball and a dimpled one

havior does not depend directly on the actual value of the viscosity, which may in fact be low. Under the circumstances of the low Reynolds number the pressure difference attempting to propel a parcel of the fluid is almost matched by the opposing viscous forces, with the result that acceleration is minimal.

For a golf ball the Reynolds number is in a middle range, high enough to make separation and wake turbulence likely but not high enough to create a turbulent boundary layer. Dimpling the surface (even at the cost of increased skin-friction drag) is justified. What about a boomerang? Unfortunately the Reynolds number of a typical boomerang also lies in a middle range of values, making the effect of roughening unclear. If the performance of a particular boomerang is not much affected by early separation of the boundary layer, roughening the surface is inefficient because the overall drag is increased by the additional drag of skin friction. On the other hand, if the performance of a boomerang is considerably affected by early separation, a roughening of the surface may increase its flight time.

A related consideration in reducing the chance of separation is whether or not the object should be streamlined. Airplane wings are streamlined to reduce separation and the drag caused by pressure difference. Should the arms of a boomerang be streamlined too? For the sake of simplicity again consider a classic airfoil. When air in the boundary layer moves over the top and toward the rear, it meets a progressive increase in pressure that threatens to stop its progress and force a separation. If the path to the rear is short, the pressure increase is sudden and separation may be unavoidable. Hence an object such as an airplane wing has a tapered rear to make the pressure increase gradual. The tapering also increases the distance over which the passing air rubs against the surface, thereby increasing the skin-friction drag. Still, the avoidance of separation is worth the increased drag.

Shapiro gives an example of streamlining in his book *Shape and Flow: The Fluid Dynamics of Drag*. He compared the overall drag on a cylindrical wire and a streamlined airfoil when air passed them at 210 miles per hour. The wire had the same drag as the airfoil, which was almost 10 times the width of the wire. Intuition suggests that the much thicker airfoil should have the greater drag, but intuition does not always take into account the subtle effects of boundary-layer separation and pressure-difference drag. At high Reynolds numbers it is better to streamline the object.

Suppose the fluid passes the object at a Reynolds number that is low enough to make separation unlikely. Streamlin-

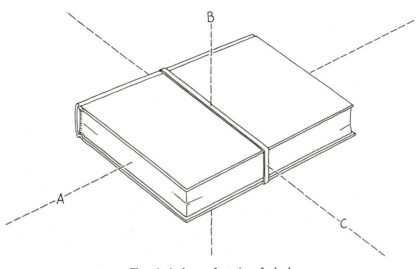

The principal axes of rotation of a book

ing the object (at the cost of increased skin-friction drag) would then be wrong. Instead the object should be blunt, presenting as little surface as possible for the air to pass.

Should the arms of a boomerang be streamlined? They probably should be, but I have not found conclusive evidence to verify my guess. Once again the Reynolds number for a typical boomerang is in the awkward middle region, and I do not know if separation is sufficiently common to justify the increased skin-friction drag the boomerang would have if the arms were streamlined.

You may wonder why none of the boomerang shapes is just a straight stick. Even if a straight stick is suitably curved on one side and flattened on the other so that it presents a classic airfoil to the passing air, it will not serve as a boomerang. The reason is a bit subtle, because it involves the stability of a rotating object against the small perturbations it encounters as it rotates.

Suppose you bind a book with a strong rubber band so that the book remains closed and flip it into the air. As is shown in the illustration above, the book can rotate about three principal axes. Rotation about two of them (either *A* or *B*) is stable, but around the third axis (*C*) the book wobbles vigorously.

The axes are characterized by the distribution of the book's mass with respect to an axis. With rotation about axis *A* the book has its mass distributed as close to the axis of rotation as possible. The book's moment of inertia is the lowest for this axis. The moment of inertia is the highest for axis *B* because then the book has its mass distributed as far as possible from the axis. About axis *C* the moment of inertia is intermediate and the rotation of the book is unstable.

The stability depends on what the angular momentum of the book does when it deviates slightly from being ex-

actly parallel to the principal axis about which the book is rotating. (The book's angular momentum is a vector with a magnitude that is the spin rate multiplied by the moment of inertia and with a vector direction that is perpendicular to the plane in which the book is spinning.) For two choices of a principal axis (*A* and *B*) a wandering angular-momentum vector is quickly rotated (precessed) back to being parallel to the principal axis. For axis *C* a wandering angular-momentum vector is rotated away from being parallel. If you could carefully spin the book around the wobbly axis with the angular-momentum vector exactly parallel to the axis, the book would spin stably. Any deviation from that parallel condition (a likely development) will increase quickly during the flight and the book will be unstable. If the greatest moment of inertia and the intermediate one are almost equal, as they would be for a square book, the spins around both of the axes associated with them will be unstable.

The standard banana-shaped boomerang spins around the axis about which it has the greatest moment of inertia, and so its spin and flight are stable. With a straight boomerang, however, the greatest moment of inertia and the intermediate one would be almost equal, and the boomerang would be unstable. The wobbling would prevent the boomerang from meeting the air at the proper angle of attack, and so the device would get no lift. In short, a straight boomerang does not boomerang.

A curious feature of a regular boomerang is the tendency for it to "lie down," that is, for its spin plane to rotate from being almost vertical to being horizontal. At the beginning of a flight the boomerang's forward velocity is great enough so that the spin plane needs only a slight tilt from the vertical to have enough upward lift to counter the weight of the boomerang. As air drag slows the forward velocity, progressively more of the lift has to be upward to hold the boomerang up. By the time the boomerang returns to the thrower the spin plane is almost horizontal.

Two possible causes for lying down can be identified. One, which is true for all types of boomerang, involves the air deflected by the arm that is turning forward of the boomerang's center of mass. Whenever an arm rotates through the forward position, it deflects the passing airstream, which (slightly later) flows around the other arm as the boomerang travels forward. Hence the trailing arm does not meet undisturbed air and does not experience the same lift as the leading arm.

For example, just after a throw by a right-handed thrower the boomerang has a horizontal lift to the left of the thrower. When the air leaves the leading arm, it must be deflected to the right of the thrower. (According to Newton's laws of motion, every action has an equal and opposite reaction. If the boomerang is forced to the left by the air, the air must be forced to the right by an equal amount.) As the trailing arm turns into air that is already flowing to the right it has a bit less lift than the leading arm because the air does not flow past the leading arm at the best attack angle. (The attack angle may even be negative.) The result is that as a boomerang spins, the leading arm always has a bit more lift than the trailing arm.

The difference in forces attempts to tilt the spin plane so that the rear of the boomerang would swing to the right (as viewed by the thrower) and the front would swing to the left. As I argued last month, this simple tilting does not ap-

pear because the boomerang is spinning and so has angular momentum. Instead what happens is that the torque resulting from the forces causes the spin plane to precess, that is, to rotate about a horizontal axis until the spin plane is horizontal by the end of the flight.

To make sense of the lying down imagine an overhead view of a right-handed boomerang that you have just launched with its spin plane vertical and therefore with its angular-momentum vector to your left. On the average the lift in front of the center of mass is more than the lift behind the center of mass because of the deflected air met by the rear of the boomerang. For the sake of simplicity consider the lift on the rear to be zero. (The results of my argument are the same even if you give that lift its proper value.) The change in angular momentum is due to the torque caused

How a boomerang "lies down" in flight

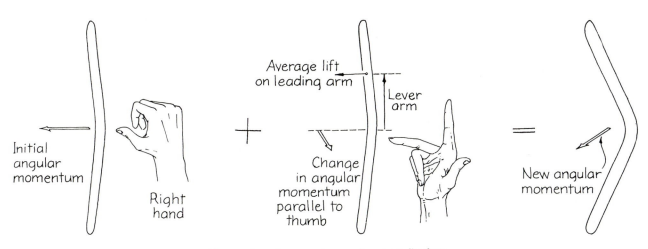

Change of angular momentum as a boomerang lies down

by the average lift on the forward section. To locate the direction of the change you should employ a rule I gave last month: Point the index finger of your right hand from the center of the boomerang toward the place where lift is applied on the boomerang's arm, point your middle finger parallel to the lift and then stretch your thumb perpendicularly to both fingers. Your thumb shows you the direction of the change in angular momentum: it is upward, perpendicular to the existing angular momentum.

Since the initial angular momentum is to the left and the change is upward, the angular-momentum vector rotates to be more upward, with the result that the spin plane of the boomerang is also rotated, turning from the vertical toward the horizontal. The boomerang has begun to lie down. This rotation, called precession because it is a rotation of the angular-momentum vector, continues throughout the flight. By the time the boomerang has returned to the thrower the spin plane is almost horizontal and the lift on the boomerang is almost entirely upward.

The second possible cause of lying down is similar. Perhaps one arm has a shape that gives more lift than the other arm. (In the illustration on the bottom of this page the arm with more lift is labeled A and the other arm is labeled B.) Both arms have the most lift when they move into the upward position (as I noted last month) because the air speed past the arm is highest when the arm is turning into the oncoming airstream. When A is in its best position, it happens to lie in front of the center of mass. When B is in its best position, it is behind the center of mass. If the greatest lift exerted on A is always greater than that exerted on B, as will be determined by the cross-sectional shape of the arms, the lift on the average will be greater in front of the center of mass than it is behind it. Once again a difference between the force in

front of the center of mass and the force behind it creates a torque that tilts the spin plane. Just as in the argument about the deflected air, the torque rotates the angular-momentum vector of the boomerang from being initially horizontal to being finally upward, causing the boomerang to lie down.

If the boomerang did not lie down, its flight would be shorter and far less interesting, because at some point during the flight the forward velocity would be too low to provide enough upward lift to support the boomerang. If a boomerang lies down so quickly that it is horizontal about halfway through the flight, it might continue to precess, swinging its angular-momentum vector out of the vertical and thereby dipping its spin plane below the horizontal. The boomerang would then begin to curve in the opposite direction and might even follow a figure-eight path on its return to the thrower.

It takes as much effort to build a boomerang that flies straight as it does to build one that returns. The horizontal lift must be almost eliminated, but enough vertical lift must remain to hold the boomerang up against its own

weight. The horizontal lift is removed by twisting the arms to give a negative angle of attack near the tips of the arms and a positive angle of attack near the center. The overall lift is greatly reduced, but not quite to zero. The boomerang is thrown with its spin plane nearly horizontal, so that the net lift is almost vertical.

If the boomerang is launched with its spin plane about 20 degrees above the horizontal, a small horizontal lift causes it to fly in a straight line toward the thrower's left. The small amount of lying down makes the spin plane rotate until it is horizontal and then makes it dip below the horizontal. Afterward the boomerang veers somewhat to the thrower's right. Since the path is not perfectly straight, the thrower must know the boomerang well if a distant target is to be hit.

Explanations of the flight and return of boomerangs have been put forward for many years, but much of the early work was erroneous and lacking in experimental verification. Except for a modern understanding of airfoil theory, the basic mechanics of boomerang flight were known as early as 1837, but appar-

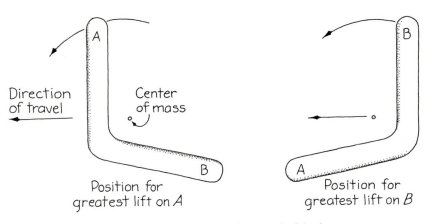

Positions of maximum lift on the respective arms of a flying boomerang

ently the information was ignored or misunderstood by later writers. A classic theoretical explanation was given by G. T. Walker in his article "On Boomerangs" in the *Philosophical Transactions of the Royal Society of London* in 1897.

The most exhaustive modern treatment of the subject of boomerangs is the doctoral dissertation by Felix Hess listed in the bibliography for this month [*page 67*]. Hess reviews virtually all the older information and then examines boomerangs himself both experimentally and theoretically. Some of his earlier information was published in this magazine [see "The Aerodynamics of Boomerangs," by Felix Hess; SCIENTIFIC AMERICAN, November, 1968]. Much more information is available in the dissertation, including several interesting three-dimensional illustrations of computer-simulated flights that the reader examines through a stereoscope viewer provided in the back of the book.

A certain amount of research on boomerangs is carried out in wind tunnels in which investigators study the paths of airflow around a stationary boomerang. Much can still be done by amateurs throwing carefully constructed boomerangs. If you pursue this research, your first problem will be to determine what your boomerang does during the flight, because you will be at only one end of the flight and can easily misjudge distance and height. To determine the distance you can station several observers along the flight path. To determine the height you can measure the angle the path makes in your field of view. By knowing both the distance and the angle you can calculate the height the boomerang attains.

Some investigators have chosen to attach a light to the boomerang and throw it in darkness while leaving the aperture of a camera open. The result is a trail of light on the photograph marking the boomerang's path. Examples of the technique were shown in Hess's article in this magazine. With a single camera, however, the three-dimensional nature of the path is difficult to appreciate. Even if you position two cameras to photograph the path from different perspectives, you will still have trouble combining their separate information to get the three-dimensional trajectory. Better results might be obtained if you photograph the lighted boomerang with the stereoscopic technique I described in this department last December. When the resulting two photographs are put in a stereoscope, one sees a combined image that has depth.

Hess and others have lighted boomerangs by attaching a small circuit composed of batteries, an electric oscillator and a small light. The arrangement is convenient, but unless you sink the objects into the wood (as Hess did) it has the disadvantage of distorting the airflow around the boomerang. The batteries and the oscillator are fastened to the center of the boomerang with only the small light bulb out on an arm, so that the boomerang's moment of inertia is altered as little as possible. A sparkler of the kind that celebrators light on the Fourth of July would also show the flight path.

Many variables stand ready for investigation if you would like to experiment with boomerangs. Although Hess has done a marvelous job in simulating boomerang flight with his computer programs, even he has not ascertained the variables to the point of designing the best boomerang or turning a bad one into a good one by some small change in construction. The major variables are those of the launching and those of the construction. A boomerang you bought would serve for a study of the variables in launching, but you would have to build one to examine the variables in construction.

To investigate launching try to find a large, open space with little wind. Choose one variable and then keep all the others constant as closely as you can. I know that holding the launching conditions constant is difficult because you will never make two successive tosses in exactly the same way. Photographing the launching with a motion-picture camera would enable you to objectively compare the launch angle, the spin and the forward velocity from one trial to another. How do the flight time, the total distance traveled and the maximum height depend on the direction of launching (normally you throw toward the horizon), the angle between the spin plane and the vertical (the greater the angle, the steeper the climb), the spin rate and the forward velocity?

With your homemade boomerangs how does the flight depend on the cross-sectional shape of the arms, the amount of twisting and the weight? I would be particularly interested to know if streamlining the arms is always advantageous. Should the tips of the arms be tapered or rounded? Should the boomerang be narrow in the center and wider at the tips of the arms? Does the flight change if the leading edge is sharp instead of blunt?

Since the effect of surface roughness is unresolved, you might want to explore it. Instead of cutting holes or grooves in a wood boomerang, try attaching cellophane tape that is sticky on each side. The tape itself might create a turbulent boundary layer, or you could create turbulence by sprinkling sand on the sticky stuff. Some researchers have investigated the effect of turbulence by mounting a thin wire just in front of the leading edge of each arm. Hess's results in such an experiment were inconclusive.

Some boomerang enthusiasts have discovered that by adding ballast to the arms near the tips they can greatly increase the distance of the outward flight of a returning boomerang. In 1972 Herb A. Smith, whose "Gem" design for a boomerang I described last month, threw such a boomerang for an outward distance of 108 yards. According to the *Guinness Book of World Records,* that is the longest outward flight ever recorded. Smith suggests adding ballast amounting to as much as a third of the boomerang's weight and countersinking the ballast (usually pieces of lead) in holes drilled about an inch from the tip of an arm. Weight added at the center will probably keep the boomerang in a more level flight, but more distance might be achieved with the ballast near the tips because then the boomerang's moment of inertia is also increased. You might see how much effect ballast has on your homemade boomerangs. If world records interest you, you might try to beat Smith's throw.

NOTES

The United States Boomerang Association publishes a newsletter titled "Return Mail and Many Happy Returns" that is chock-full of information on boomerang events, designs, and records. (The address of the association is in the bibliography.) The latest issue (fall 1984) lists four new records. Rob Croll now holds the record of 653 consecutive catches. His personally designed boomerang was an acute-angle V shape. To qualify for this record, one must throw the boomerang out for at least 20 meters and then catch it on its return. Croll tossed for over three hours. He knew his boomerang's flight so well that even when he was too fatigued to follow the flight, he could still catch the boomerang by putting his hands in the correct spot.

Mike Forrester holds the current record for keeping a boomerang in the air. To qualify for this record, one must also catch the boomerang. Forrester's boomerang was aloft for 50.8 seconds. (Peter Ruhf once had a practice throw in which his boomerang was aloft for one minute and 35 seconds, after which he caught it.) The fast-catch contest requires that the boomerang thrower execute five rapid tosses (for a distance of at least 20 meters) with catches. The record for the least time in this contest is now held by John Flynn, who performed the task in 21.8 seconds. He has since had a practice run of 19.8 seconds. Barney Ruhe holds the record for consecutive juggling catches. He managed 161 catches, more than topping the previous record of 69.

The grandest boomerang contest is the one for greatest distance. The record is now held by Ruhf, who tossed a boomerang 375 feet in a contest conducted at Randwick, Sydney, Australia, on June 28, 1982. Benjamin Ruhe, who has long been active in the sport of boomeranging, advises me that this record will soon fall because of new boomerangs that are being fashioned out of aluminum. These boomerangs have large moments of inertia because of the metal's large mass. When they are launched with a strong flip of the wrist, they have large rotational kinetic energies that keep them aloft for a long time. However, these boomerangs are highly dangerous. Ruhe describes them as "flying knives" and urges caution about throwing them when other people are around. They also should be caught with caution and always with heavy gloves.

Ruhe has also told me about a neat feature added to some boomerangs. They are fitted with "cold-light" tubes that emit light for several hours through chemical reactions. Such boomerangs can be flown at night, allowing the boomerang thrower to enjoy empty playing fields and low winds.

Ruhe's excellent book is now sold in paperback under the new title *Boomerang*. (See the bibliograhy for his address, from which you can order the book.) Ruhe also publishes a boomerang catalog, from which you can order an assortment of fine boomerangs, including the night fliers. Wham-O no longer manufactures the boomerang I described in the first of my boomerang articles. The company now sells a tri-blader called the Wham-O-Rang. This rubber-tipped boomerang travels out 30 or 40 yards and then returns.

The Smithsonian no longer conducts its annual boomerang class and tournament. If you are interested in contests, consult the newsletter from the United States Boomerang Association. The association's 1985 National Championship is scheduled to be held July 19 through 21 at the California State University at Northridge, which is near Burbank.

Many people corrected my statement that a straight length of wood cannot be thrown so as to return to the thrower. Bernard Mason's book (listed in the bibliography) contains plans for constructing such a stick, which he calls a tumblestick. Recently Henk Vos has submitted a paper to the *American Journal of Physics* in which he analyzes the flight of this type of boomerang.

A. J. Cochran of the University of Aston in Birmingham, England, corrected my analysis of how dimples aid the flight of a golf ball and thus might influence the flight of a boomerang. "It is true that the addition of dimples or other roughening reduces the drag on smooth spheres, at least at certain velocities, and that the reason for this is the one given: transition from a laminar to a turbulent boundary layer, with consequent delayed separation of it, and narrowing of the wake.

"The main benefit of the dimpling in a golf ball is to help give it lift. In all properly struck golf shots, spin is imparted to the ball about a horizontal axis at right angles to the intended direction of flight—"backspin" as it is known. The effect of this is to speed up flow over the top of the ball and slow it under the bottom, with the consequent pressure difference (low above, high below) creating a lift force. This effect is the well known Magnus effect, and it operates in many other ball games—tennis, baseball, soccer, table tennis.

"Dimpling is more effective in encouraging the asymmetric flow than a smooth surface, and the lift produced can be as great as the weight of the ball when at the beginning of the golf shot the spin is 3,000 to 8,000 revolutions per minute. The great length of a good golf shot arises from the fact that the ball 'glides' in this way. The dimpling on a golf ball is therefore designed to optimise lift and drag properties with the former being more important."

Several people have experimented with boomerangs having rough surfaces. James W. Boyd, Sr., of Columbus, Ohio, told me how he ruffled the surface of a plastic boomerang with dabs of airplane glue. The boomerang seemed to fly unchanged. John B. Mauro of Richmond, Virginia, conducted three experiments. He first constructed an L-shaped boomerang from ten-ply birch and painted and lacquered it for smoothness. It flew about 20 to 25 meters before executing a turn. A forceful throw was needed before the boomerang would return. When he roughened the top of the boomerang with coarse sandpaper, it "became more manageable, flew a shorter range, and had better lying-down qualities." When he then sanded it with fine sandpaper, the original flight characteristics returned.

He next gouged dimples into the top of the boomerang. After painting and lacquering it again, he found that the flight path was "substantially shorter, bringing about a faster, closer return with less throwing effort. The boomerang's lying down improved too. It seems that dimples produced additional lift as did general roughening.

"A similar experiment was conducted on an omega-shaped boomerang using another airfoil design to reduce lift. In this case, from the center of each arm to the elbow, each arm was given trailing edges so that only the outer half of the arms had the conventional boomerang airfoil shape. This design considerably reduced the boomerang's lift area. It flew low and level and had a tendency to hit the ground about halfway through the flight unless thrown very hard with precisely the appropriate angle.

"As before, the boomerang was lacquered for smoothness. Oblong dimples were gouged on the upper surface, but this time only on the leading arm. The trailing arm remained smooth. The flight path remained low, that is, the boomerang did not increase altitude upon its return as is normal for boomerangs. However, its distance was reduced and its lying down improved, and it took a less aggressive launch.

"A boomerang designed in the shape of the ordinates of the normal probability curve had very little lift. About three fourths of the way around the flight it would dive to the ground. To get it to return to the thrower required a very strong and precise throw. Often, even with a strong throw, it would dive to the ground, so much so that in one instance it broke."

According to Mauro, after gouges were put along the top of the leading arm three-fourths of the way from the tip of the arm to the elbow, the boomerang's flight became quite stable and required moderate force in its launch. "The radius seemed to be shorter. When similar gouges were put on the top of the trailing arm, the boomerang's flight remained stable. The only noticeable difference was that it climbed much higher on its return."

Robert L. Easton, a teacher at Red Lion Area Senior High School in Red Lion, Pennsylvania, described to me how he has introduced boomerang flight to his students, taking them from the simple designs, through the multiple-bladed ones, and finally to the large hooks and weighted varieties. Easton, who has now made hundreds of boomerangs, offers the following tips: "I underestimated the importance of undercutting the lift arm out near the tip. As a result my first boomerangs did not lie down but returned still in a nearly vertical plane and I couldn't catch them. Also some of my first efforts were too rounded at the leading edge and returned with too much spin left. It took a lot of courage to catch them. I noticed too that my V-shaped boomerangs were far more sensitive to slight differences in the leading edge than were my U-shaped models.

"I have just begun experimenting with weighted models. I considered solid solder rather than lead. The lower density means using more weights but solder is very easy to work with and does not form the familiar gray oxide.

"I don't have access to a wind tunnel but I've been experimenting with the lift of various cross sections using nothing but a shop vacuum and a triple-beam balance. I glue the wing section to a pedestal sitting on the balance pan and measure the apparent weight loss when a stream of air blows over the wing. In this way I learned that it definitely pays to sand and varnish the boomerang.

"One reason boomerangs have not evolved faster, I suspect, is the very low profit in making them and the resultant reluctance of individuals to share their discoveries. I was practically sworn to secrecy once when another maker shared his observations with me."

Jürgen Krüger of the Klinikum der Albert-Ludwigs-Universität in Freiburg, West Germany, commented on my observation that there is no way to remedy a persistently unsuccessful boomerang. "I largely agree with that statement with one exception. Ideally, the end of the pear shaped path should be straight (as seem from above) and directly towards the thrower. Only then will there be a brief stop of the forward motion (and an increase in rotational speed) so the boomerang can be caught. However, if the total path is more of a spiral so that the boomerang, near the end of its flight, moves from left to right in front of the thrower, then the total surface of the boomerang should be increased by a flat piece of cardboard glued to the bottom of the boomerang, protruding preferentially into the inner side of the boomerang's angle. The cardboard will reduce the anticlockwise curl of the final path and, with larger added surfaces, even produce a clockwise curl. Of course, the opposite phenomenon is remedied by a reduction of surface near the bend of the boomerang where the airfoil function is not so important.

"The simplified explanation is that the boomerang behaves like a flat disk. In a situation like the one depicted in your article the air tends to put the entire boomerang transverse to the forward direction, but as it spins, the result of this torque is that the angle of launch increases during the flight. If the boomerang is too slim, there is not enough surface so that the torque is too small. As a consequence, the angle of launch is nearly conserved near the end of the path, and so the boomerang continues to move on a curve. Only when the angle of launch has increased to a right angle (that is, the boomerang rotates horizontally) will the projection of the path on the ground be straight."

Many other people wrote me about their experiments with boomerangs. Indeed, some people have written dissertations on the subject. If you investigate boomerangs either theoretically or experimentally, I would enjoy hearing about your findings.

BIBLIOGRAPHY

Literature search conducted through February 1985.

1. AMUSEMENT PARK PHYSICS

HARRY G. TRAVER: LEGENDS OF TERROR. Richard Munch. Amusement Park Books, 1982.

ANOTHER LOOK AT THE UNIFORM ROPE SLIDING OVER THE EDGE OF A SMOOTH TABLE. Domingo Prato and Reinaldo J. Gleiser in *American Journal of Physics*, Vol. 50, No. 6, pages 536–539; June, 1982.

COMMENT ON "ANOTHER LOOK AT THE UNIFORM ROPE SLIDING OVER THE EDGE OF A SMOOTH TABLE." J. R. Sanmartin and M. A. Vallejo in *American Journal of Physics*, Vol. 51, No. 7, page 585; July, 1983.

2. RACQUETBALL

MATRICES AND SUPERBALLS. George L. Strobel in *American Journal of Physics*, Vol. 36, pages 834–837; 1968.

KINEMATICS OF AN ULTRAELASTIC ROUGH BALL. Richard L. Garwin in *American Journal of Physics*, Vol. 37, pages 88–92; 1969.

THE COMPLETE BOOK OF RACQUETBALL. Steve Keeley. DBI Books, 1976.

HANDBALL BASICS. George J. Zafferano. Sterling Publishing Co., 1977.

ADVANCED RACQUETBALL. Steve Strandemo and Bill Burns. Simon and Schuster, 1981.

THE DYNAMICS OF SPORTS: WHY THAT'S THE WAY THE BALL BOUNCES, 2d ed. David F. Griffing. The Dalog Company, P. O. Box 243, Oxford, Ohio 45056; 1982.

THAT'S HOW THE BALL BOUNCES. Howard Brody in *The Physics Teacher*, Vol. 22, No. 8, pages 494–497; November, 1984.

3. BILLIARDS AND POOL

BILLIARD-BALL COLLISION EXPERIMENT. Jane Higginbotham Bayes and William T. Scott in *American Journal of Physics*, Vol. 31, pages 197–200; 1963.

MECHANICS. Arnold Sommerfeld. Academic Press, 1964, pages 158–161, 250–251.

CLASSICAL MECHANICS: A MODERN PERSPECTIVE. V. Barger and M. Olsson. McGraw-Hill Book Company, 1973, pages 186–188.

MASTERING POOL. George Fels. Contemporary Books, 1977.

BYRNE'S TREASURY OF TRICK SHOTS IN POOL AND BILLIARDS. Robert Byrne. Harcourt Brace Jovanovich, Publishers, 1982.

THE DYNAMICS OF SPORTS: WHY THAT'S THE WAY THE BALL BOUNCES, 2d ed. David F. Griffing. The Dalog Company, P. O. Box 243, Oxford, Ohio 45056, 1982.

4. MARTIAL ARTS

JUDO ON THE GROUND. E. J. Harrison. W. Foulsham & Co., 1954.

AIKIDO AND THE DYNAMIC SPHERE: AN ILLUSTRATED INTRODUCTION. A. Westbrook and O. Ratti. Charles E. Tuttle Co., 1970.

TRADITIONAL AIKIDO: SWORD, STICK AND BODY ARTS. Morihiro Saito. Minato Research & Publishing Co., 1973–1974.

5. BALLET

AN ANALYSIS OF TURNS IN DANCE. K. L. Laws in *Dance Research Journal*, No. 11–12, page 16; 1978–1979.

PHYSICS AND BALLET: A NEW PAS DE DEUX. Kenneth Laws in *New Directions in Dance*, edited by Diana Theodores Taplin. Pergamon Press, 1979.

PRECARIOUS AURORA—AN EXAMPLE OF PHYSICS IN PARTNERING. Kenneth Laws in *Kinesiology for Dance*, No. 12, pages 2–3; August, 1980.

THE PHYSICS OF DANCE. Kenneth Laws. Schirmer Books, 1984.

THE PHYSICS OF DANCE. Kenneth Laws in *Physics Today*, Vol. 38, No. 2, pages 24–31; February, 1985.

6. RATTLEBACKS

DYNAMICS OF A SYSTEM OF RIGID BODIES, ADVANCED PART. E. J. Routh. Macmillan, 1930, pages 204–205.

A TREATISE ON GYROSTATICS AND ROTATIONAL MOTION: THEORY AND APPLICATION. Andrew Gray. Dover Publications, 1959, pages 363–366.

AN ELEMENTARY TREATMENT OF THE THEORY OF SPINNING TOPS AND GYROSCOPIC MOTION. Harold Crabtree. Chelsea, 1967, pages 7, 54.

A MATHEMATICAL MODEL OF THE "RATTLEBACK." T. K. Caughey in *International Journal of Non-Linear Mechanics*, Vol. 15, No. 4/5, pages 293–302; 1980.

REALISTIC MATHEMATICAL MODELING OF THE RATTLEBACK. T. R. Kane and D. A. Levinson in *International Journal of Non-Linear Mechanics*, Vol. 17, No. 3, pages 175–186; 1982.

6 and 7. TIPPE TOPS

THE RISING TOP, EXPERIMENTAL EVIDENCE AND THEORY. A. D. Fokker in *Physica*, Vol. 8, pages 591–596; 1941.

NOTE ON THE BEHAVIOR OF A CERTAIN SYMMETRICAL TOP. J. A. Jacobs in *American Journal of Physics*, Vol. 20, pages 517–518; 1952.

ON A CASE OF INSTABILITY PRODUCED BY ROTATION. J. L. Synge in *Philosophical Magazine*, Series 7, Vol. 43, pages 724–728; 1952.

ON THE INFLUENCE OF FRICTION ON THE MOTION OF A TOP. C. M. Braams in *Physica*, Vol. 18, pages 503–514; 1952.

ON TOPS RISING BY FRICTION. N. M. Hugenholtz in *Physica*, Vol. 18, pages 515–527; 1952.

THE TIPPE TOP (TOPSY-TURVY TOP). W. A. Pliskin in *American Journal of Physics*, Vol. 22, pages 28–32; 1954.

THE TIPPE TOP. C. M. Braams in *American Journal of Physics*, Vol. 22, page 568; 1954.

TIPPE TOP (TOPSY-TURNEE TOP) CONTINUED. A. R. Del Camp in *American Journal of Physics*, Vol. 23, pages 544–545; 1955.

THE TIPPE TOP AGAIN. I. M. Freeman in *American Journal of Physics*, Vol. 24, page 178; 1956.

SPINNING TOPS AND GYROSCOPIC MOTIONS. J. Perry. Dover Publicatons, 1957, pages 39–57.

ANGULAR MOMENTUM AND TIPPE TOP. J. B. Hart in *American Journal of Physics*, Vol. 27, page 189; 1959.

THE TIPPY TOP. F. Johnson in *American Journal of Physics*, Vol. 28, pages 406–407; 1960.

AN ELEMENTARY TREATMENT OF THE THEORY OF SPINNING TOPS AND GYROSCOPIC MOTION. Harold Crabtree. Chelsea, 1967, pages 5–6, 51, 155.

THE TIPPE TOP REVISITED. Richard J. Cohen in *American Journal of Physics*, Vol. 45, No. 1, pages 12–17; January, 1977.

ELEMENTARY DYNAMICS OF SIMPLE MECHANICAL TOYS. W. Bürger in *Mitteilungen der Gesellschaft für Angewandte Mathematik und Mechanik*, Vol. 2, pages 21–60; July, 1980.

7. TOPS

THEORETICAL MECHANICS: AN INTRODUCTION TO MATHEMATICAL PHYSICS. Joseph Sweetman Ames and Francis D. Murnaghan. Ginn and Company, 1929, pages 231–255.

INTRODUCTION TO THEORETICAL PHYSICS, 2d ed. Leigh Page. D. Van Nostrand Company, 1935.

SPINNING TOPS AND GYROSCOPIC MOTIONS. J. Perry. Dover Publications, 1957.

ON THE STABILITY OF A SPINNING TOP CONTAINING LIQUID. K. Stewartson in *Journal of Fluid Mechanics*, Vol. 5, pages 577–592; 1959.

MECHANICS OF THE GYROSCOPE: THE DYNAMICS OF ROTATION. Richard F. Deimel. Dover Publications, 1960, pages 75–106.

DYNAMICS OF RIGID BODIES. William Duncan MacMillan. Dover Publications, 1960, pages 239–249.

THE DYNAMICS OF PARTICLES AND OF RIGID, ELASTIC, AND FLUID BODIES. Arthur Gordon Webster. Dover Publications, 1960, pages 275–316.

ON THE PRECESSION AND NUTATION OF GYROSCOPES. H. L. Armstrong in *American Journal of Physics*, Vol. 35, pages 883–885; 1967.

THE TOP: UNIVERSAL TOY, ENDURING PASTIME. D. W. Gould. Clarkson N. Potter, 1973.

THE GYROSCOPE: AN ELEMENTARY DISCUSSION OF A CHILD'S TOY. William Case in *American Journal of Physics*, Vol. 45, No. 11, pages 1107–1109; November, 1977.

A PHYSICAL EXPLANATION OF THE GYROSCOPE EFFECT. P. L. Edwards in *American Journal of Physics*, Vol. 45, No. 12, pages 1194–1195; December, 1977.

A REALISTIC SOLUTION OF THE SYMMETRIC TOP PROBLEM. T. R. Kane and D. A. Levinson in *Journal of Applied Mechanics*, Vol. 45, No. 4, pages 903–909; December, 1978.

BEHAVIOR OF A REAL TOP. Ledo Stefaini in *American Journal of Physics*, Vol. 47, No. 4, pages 346–350; April, 1979.

PRECESSION OF A GYROSCOPE. J. Higbie in *The Physics Teacher*, Vol. 19, page 210; March, 1980.

TOPS—WITH STRINGS ATTACHED. M. J. Lea in *Physics Education*, Vol. 17, pages 24–25; 1982.

THE YO-YO TOP. Richard Harding and Robert Prigo in *The Physics Teacher*, Vol. 22, No. 1, pages 36–37; January, 1984.

8. WOBBLING

COIN SPINNING ON A TABLE. M. G. Olsson in *American Journal of Physics*, Vol. 40, pages 1543–1545; October, 1972.

SPINNING OBJECTS ON HORIZONTAL PLANES. Lorne A. Whitehead and Frank L. Curzon in *American Journal of Physics*, Vol. 51, No. 5, pages 449–452; May, 1983.

9 and 10. BOOMERANGS AND FRISBEES

ON BOOMERANGS. G. T. Walker in *Philosophical Transactions of the Royal Society of London*, Series A, Vol. 190, pages 23–41; 1898.

A TREATISE ON GYROSTATICS AND ROTATIONAL MOTION: THEORY AND APPLICATION. Andrew Gray. Dover Publications, 1959, pages 291–296.

THE AERODYNAMICS OF BOOMERANGS. Felix Hess in *Scientific American*, Vol. 219, No. 5, pages 124–136; November, 1968.

MANY HAPPY RETURNS. Peter Musgrove in *New Scientist*, Vol. 61, No. 882, pages 186–189; January 24, 1974.

BOOMERANGS: MAKING AND THROWING THEM. Herb A. Smith. Gemstar Publications, 1975.

PROJECT BOOMERANG. Allen L. King in *American Journal of Physics*, Vol. 43, No. 9, pages 770–773; September, 1975.

THE PHYSICS OF FRISBEE FLIGHT. Jay Shelton in *Frisbee: A Practitioner's Manual and Definitive Treatise*, edited by Stancil E. D. Johnson. Workman Publishing Co., 1975.

BOOMERANGS, AERODYNAMICS, AND MOTION. Felix Hess. 1975 dissertation available from the author, c/o Dr. H. Rollema, Eikenlaan 51, Peize, Netherlands.

WHY BOOMERANGS RETURN. David Robson, available from the author, 4602 Schenley Road, Baltimore, MD 21210; 1977.

THE BOOMERANG. M. J. Hanson in *The School Science Review*, pages 428–437; March, 1977.

BOOMERANGS: HOW TO MAKE AND THROW THEM. Bernard S. Mason. Dover Publications, 1974.

MANY HAPPY RETURNS: THE ART AND SPORT OF BOOMERANGING. Benjamin Ruhe. The Viking Press, 1977.

BOOMERANG. Benjamin Ruhe. Minner Press, 1982; available from the author, P. O. Box 7324, Washington, DC 20044.

WHY BOOMERANGS BOOMERANG (AND KILLING STICKS DON'T). Jacques Thomas in *New Scientist*, Vol. 99, pages 838–843; September 22, 1983.

THE STRAIGHT BOOMERANG OF BALSA WOOD. Henk Vos, submitted to *American Journal of Physics*.

RETURN MAIL AND MANY HAPPY RETURNS, a newsletter published by the United States Boomerang Association, 4030–9 Forest Hill Avenue, Richmond, VA 23225.

11. OTHER ROTATION SUBJECTS

YO-YO TECHNICS IN TEACHING KINEMATICS. I. L. Kofsky in *American Journal of Physics*, Vol. 19, pages 126–127; 1951.

EFFECT OF SPIN AND SPEED ON THE LATERAL DEFLECTION (CURVE) OF A BASEBALL; AND THE MAGNUS EFFECT FOR SMOOTH SPHERES. L. J. Briggs in *American Journal of Physics*, Vol. 27, pages 589–596; 1959.

HULA-HOOP: AN EXAMPLE OF HETEROPARAMETRIC EXCITATION. T. K. Caughey in *American Journal of Physics*, Vol. 28, pages 104–109; 1960.

COWBOY ROPING AND ROPE TRICKS. C. Byers. Dover Publications, 1966.

PUMPING ON A SWING. P. L. Tea, Jr., and H. Falk in *American Journal of Physics*, Vol. 36, pages 1165–1166; 1968.

COMMENTS ON PUMPING ON A SWING. A. E. Siegman in *American Journal of Physics*, Vol. 37, pages 843–844; 1969.

THE CHILD'S SWING. B. F. Gore in *American Journal of Physics*, Vol. 38, page 378, 1970.

MORE ON PUMPING A SWING. J. A. Burns in *American Journal of Physics*, Vol. 38, pages 920–922; 1970.

STARTING A SWING FROM REST. B. F. Gore in *American Journal of Physics*, Vol. 39, page 347; 1971.

STATIC VS. SPIN BALANCING OF AUTOMOBILE WHEELS. R. C. Smith in *American Journal of Physics*, Vol. 40, pages 199–201; 1972.

ON INITIATING MOTION IN A SWING. J. T. McMullan in *American Journal of Physics*, Vol. 40, pages 764–766; 1972.

THE PHYSICS OF SKI TURNS. J. I. Shonle and D. L. Nordick in *The Physics Teacher*, Vol. 10, pages 491–497; 1972.

AERODYNAMICS OF A KNUCKLEBALL. R. G. Watts and E. Sawyer in *American Journal of Physics*, Vol. 43, No. 11, pages 960–963; November, 1975.

MORE ON THE FALLING CHIMNEY. Albert A. Bartlett in *The Physics Teacher*, Vol. 14, No. 6, pages 351–353; September, 1976.

THE FLYING CIRCUS OF PHYSICS WITH ANSWERS. Jearl Walker. John Wiley & Sons, 1977.

THEORY OF THE CHIMNEY BREAKING WHILE FALLING. Ernest L. Madsen in *American Journal of Physics*, Vol. 45, No. 2, pages 182–184; February, 1977.

BOWLING FRAMES: PATHS OF A BOWLING BALL. D. C. Hopkins and J. D. Patterson in *American Journal of Physics*, Vol. 45, No. 3, pages 263–266; March, 1977.

LAWN BOWLS AND NEWTON'S LAW OF MOTION. T. G. L. Shirtcliffe in *Physics Education*, Vol. 14, pages 78–81; 1979.

STRANGE TO RELATE, SMOKESTACKS AND PENCIL POINTS BREAK IN THE SAME WAY. Jearl Walker in "The Amateur Scientist" of *Scientific American*, Vol. 240, No. 2, pages 158–164; February, 1979.

DO SPRINGBOARD DIVERS VIOLATE ANGULAR MOMENTUM CONSERVATION? Cliff Frohlich in *American Journal of Physics*, Vol. 47, No. 7, pages 583–592; July, 1979.

THE PHYSICS OF THE "DYNA BEE." J. Higbie in *The Physics Teacher*, Vol. 18, No. 2, pages 147–148; February, 1980.

THE PHYSICS OF SOMERSAULTING AND TWISTING. Cliff Frohlich in *Scientific American*, Vol. 242, No. 3, pages 155–164; March, 1980.

THE SPIN ON BASEBALLS OR GOLFBALLS. R. D. Edge in *The Physics Teacher*, Vol. 19, pages 308–309; April, 1980.

ELEMENTARY DYNAMICS OF SIMPLE MECHANICAL TOYS. W. Bürger in *Mitteilungen der Gesellschaft für Angewandte Mathematik und Mechanik*, Vol. 2, pages 21–60; July, 1980.

AERODYNAMIC EFFECTS ON DISCUS FLIGHT. Cliff Frohlich in *American Journal of Physics*, Vol. 49, No. 12, pages 1125–1132; December, 1981.

THE DYNAMICS OF SPORTS: WHY THAT'S THE WAY THE BALL BOUNCES, 2d ed. David F. Griffing. The Dalog Company, P. O. Box 243, Oxford, Ohio 45056; 1982.

THE YO-YO: A TOY FLYWHEEL. Wolfgang Bürger in *American Scientist*, Vol. 72, No. 2, pages 137–142; March-April, 1984.

SPORT SCIENCE: PHYSICAL LAWS AND OPTIMUM PERFORMANCE. Peter J. Brancazio. Simon and Schuster, 1984.

INDEX